Cambridge Elements ≡

Elements in the Structure and Dynamics of Complex Networks
edited by
Guido Caldarelli
Ca' Foscari University of Venice

HIGHER-ORDER NETWORKS

An Introduction to Simplicial Complexes

Ginestra Bianconi
*Queen Mary University of London and The Alan
Turing Institute*

CAMBRIDGE
UNIVERSITY PRESS

University Printing House, Cambridge CB2 8BS, United Kingdom

One Liberty Plaza, 20th Floor, New York, NY 10006, USA

477 Williamstown Road, Port Melbourne, VIC 3207, Australia

314–321, 3rd Floor, Plot 3, Splendor Forum, Jasola District Centre,
New Delhi – 110025, India

103 Penang Road, #05–06/07, Visioncrest Commercial, Singapore 238467

Cambridge University Press is part of the University of Cambridge.

It furthers the University's mission by disseminating knowledge in the pursuit of
education, learning, and research at the highest international levels of excellence.

www.cambridge.org
Information on this title: www.cambridge.org/9781108726733
DOI: 10.1017/9781108770996

First published 2021

A catalogue record for this publication is available from the British Library.

ISBN 978-1-108-72673-3 Paperback
ISSN 2516-5763 (online)
ISSN 2516-5755 (print)

Higher-Order Networks

An Introduction to Simplicial Complexes

Elements in the Structure and Dynamics of Complex Networks

DOI: 10.1017/9781108770996
First published online: November 2021

Ginestra Bianconi
Queen Mary University of London and The Alan Turing Institute
Author for correspondence: Ginestra Bianconi, ginestra.bianconi@gmail.com

Abstract: Higher-order networks describe the many-body interactions of a large variety of complex systems ranging from the the brain to collaboration networks. Simplicial complexes are generalized network structures which allow us to capture the combinatorial properties, the topology and the geometry of higher-order networks. Having been used extensively in quantum gravity to describe discrete or discretized space-time, simplicial complexes have only recently become the representation of choice for capturing the underlying network topology and geometry of complex systems.
This Element provides an in-depth introduction to the very hot topic of network theory, covering a wide range of subjects ranging from emergent hyperbolic geometry and topological data analysis to higher-order dynamics. This Element aims to demonstrate that simplicial complexes provide a very general mathematical framework to reveal how higher-order dynamics depends on simplicial network topology and geometry.

Keywords: Higher-order networks, Simplicial complexes, Higher-order dynamics, Simplicial network topology and geometry, Topological data analysis

Isbns: 9781108726733 (PB), 9781108770996 (OC)
Issns: 2516-5763 (online), 2516-5755 (print)

Contents

1 The Relevance of Higher-Order Networks
in Network Science   1

2 Combinatorial and Statistical Properties of Simplicial
Complexes   7

3 Simplicial Network Topology   30

4 Simplicial Network Geometry   46

5 Emergent Geometry   56

6 Higher-Order Dynamics: Synchronization   86

7 Higher-Order Dynamics: Percolation   99

8 Higher-Order Dynamics: Contagion Models   107

9 Outlook   113

Appendix A Maximum Entropy Ensembles of Simplicial
Complexes   115

Appendix B The Hodge Decomposition   119

Appendix C Spectral Dimension of Euclidean Lattices   121

Appendix D Topological Moves   123

Appendix E Emergent Preferential Attachment   125

Appendix F Generalized Degree Distributions of NGFs
(Neutral Model)   127

Appendix G Apollonian and Pseudo-Fractal Simplicial
Complexes   129

References   130

1 The Relevance of Higher-Order Networks in Network Science

1.1 Simplicial Complexes as Generalized Network Structures

Network science [1–4] is routed on the idea that the complexity of interacting systems can be captured by the network of interactions between their constituents. This very powerful framework has enabled the scientific community to make unprecedented progress in the understanding of complex systems ranging from the brain to society. In the last 20 years, network theory has revealed the rich interplay between the network topology and dynamics [3, 5]. It has been shown that universal statistical properties of complex networks, such as the scale-free degree distribution and the small-world nature of complex networks, are responsible for the surprising dynamical properties that processes such as percolation, epidemic spreading, Ising model and synchronization display in these networks.

Recently, mounting evidence reveals that to make the next big leap forward in understanding and predicting the behavior of complex networks it is important to abandon the framework of a simple network formed exclusively by pairwise interactions and use generalized network structures that can better capture the richness of real data. Multilayer networks [6] are well-studied generalized networks that are able to treat datasets in which interactions have different nature and connotation displaying a rich interplay between structure and dynamics. More recently the research attention has been focusing on higher-order networks [7–11] that allow capture of the many-body interactions of complex systems going beyond the pairwise interaction framework.

Consider for example three regions of the brain. These three regions can be correlated with each other pairwise via three two-body interactions or might be related by a higher-order (three-body) interaction, revealed by the fact that these three regions of the brain are typically activated at the same time. These two scenarios correspond to very different dynamics that can be distinguished only by considering higher-order networks. Indeed in the first case the higher-order network between the three regions of the brain will include just three links, while in the second case the three brain regions would form a three-body interaction indicated by a filled triangle (also called a two-dimensional simplex). In social networks a notable example of a higher-order network is constituted by the set of face-to-face interactions at a party or during a coffee break at a conference. In this context the people will form conversation groups, often involving more than two individuals, in which ideas are shared and elaborated in a way that is not reducible to a set of pairwise conversations.

Likewise in protein interaction networks proteins bind to each other forming protein complexes typically including more than two different proteins. Only when the protein complex is fully assembled is the protein complex able to perform its biological task. This indicates that the biological function of the protein complex is the result of many-body interactions between its constituent proteins and cannot be reduced to a set of pairwise interactions.

Higher-order networks fully capture the interactions between two or more nodes and are necessary to describe dynamical processes depending on many-body interactions. In recent years this research field has boomed and important new progress has been made to uncover the interplay between higher-order structure and dynamics. In this work we aim to provide a brief introduction to the subject that could be useful for graduate students and for researchers to jump start into this lively research field.

1.2 Simplicial Complexes and Hypergraphs

When faced with the problem of capturing higher-order interactions existing in a dataset, two generalized network structures are potentially useful for the researcher: simplicial complexes and hypergraphs.

Both simplicial complexes and hypergraphs capture higher-order interactions and are formed by a set of nodes $v \in \{1, 2, 3 \ldots, N\}$ and a set of many-body interactions including two or more nodes, such as

$$\alpha = [v_0, v_1, \ldots, v_d], \tag{1.1}$$

with $d \geq 1$. These many-body interactions are called *simplices* of a simplicial complex or *hyperedges* of a hypergraph. These higher-order interactions induce a very rich combinatorial structure for higher-order networks [12] that can also have very relevant consequences for higher-order dynamics, including synchronization and contagion processes [13, 14]. As both simplicial complexes and hypergraphs capture the many-body interactions in a complex system, the vast majority of many-body phenomena obtained in one framework can be directly translated into the other framework.

The only difference between simplicial complexes and hypergraphs is a subtle one: the set of simplices of a simplicial complex is closed under the inclusion of subsets of the simplices in the set while no such constraint exists for a hypergraph. This means that in a simplicial complex, if the simplex α given by

$$[v_0, v_1, v_2], \tag{1.2}$$

belongs to the simplicial complex, then the simplices

$$[v_0, v_1], \quad [v_0, v_2], \quad [v_1, v_2], \quad [v_0], \quad [v_1], \quad [v_2], \tag{1.3}$$

must also belong to the simplicial complex. In other words, if we consider a collaboration network in which three authors have written a paper together, then we should include in the simplicial complex also the three pairwise interactions between the authors and the set of the three isolated nodes. This might look like an artificial constraint, but actually it comes with the great advantage that simplicial complexes are natural topological spaces for which important topological results exist. This widely developed branch of mathematics provides a powerful resource for extracting information and revealing the interplay between topological and geometrical properties of higher-order networks and their dynamics.

The scientific research on higher-order networks is currently growing and many important results have been recently obtained in this field. In this Element, our goal is to provide a self-contained, coherent and uniform account of the results on higher-order networks. For space limitations we have chosen to focus mostly on simplicial complexes. However, on a number of occasions we will refer to results exclusively applying to hypergraphs.

1.3 A Topological Approach to Complex Interacting Systems

A simplex characterizes an interaction between two or more nodes. The simplices of a simplicial complex are glued to one another by sharing a subset of their nodes, resulting in topological spaces. Topological spaces have a number of features, for instance they can be characterized not only by the number of their connected components, like networks, but also by the number of their higher-order cavities or holes indicated by their Betti numbers. Applied topology [8, 15–19] studies the underlying topology (including the Betti numbers) of simplicial complexes coming from real data. This field has been flourishing in the last decades and was initially applied to extract information from data-clouds coming from different sources of data including, for instance, gene-expression. An important framework that has been developed in applied topology is called *persistent homology* and is based on an operation called filtration that aims at coarse-graining the data with different resolution characterizing how long topological features persist. Only recently [20, 21] has this approach been applied to real networked data and in particular to brain functional networks, which are weighted networks in which the filtration procedure is not simple coarse-graining, rather it is substituted with a change of threshold in the weights of the links. Persistence homology of complex networks is a powerful topological tool that makes extensive use of the simplicial representation of data and has shown to reveal differences not accounted for by other more traditional Network Science measures. However the possibility of using persistence homology is by no means the only benefit of using topology to analyze higher-order networks. In neuroscience [8, 22] the use of simplicial complexes has been booming in recent years and novel results show the rich interplay between topology and dynamics in the framework of the in-silico reconstruction of rat brain cortex [23]. Moreover, simplicial topology can be also used to investigate the local [24] and the meso-scale structure [25, 26] of network data.

Departing from the benefit that topology can bring to higher-order data analysis, recently it has been shown that topology, and specifically Hodge theory, can be exploited by higher-order networks for sustaining and synchronizing higher-order topological signals, i.e. dynamical variables that are not only defined on the nodes of the network, but rather like fluxes they can be defined on links or even on higher-order structures like triangles or tetrahedra [27]. Interestingly topological signals are also attracting increasing attention from the signal processing perspective [28].

This multifaceted research field clearly shows that topology is a fundamental tool to investigate higher-order network structure and dynamics.

1.4 A Geometrical Approach to Higher-Order Networks

If the links of a simplicial complex are assigned a distance, simplices have an automatic interpretation as geometrical objects, and can be understood as nodes, links, triangles, tethrahedra, etc. In particular, in absence of other data that can be used to assign a distance to each link, the network scientist can always choose to assign the same distance to each link.

Since simplicial complexes describe discrete simplicial geometries, modeling simplicial complexes opens the possibility to reveal the fundamental mechanisms of emergent simplicial geometry.

The long-standing mathematical problem of emergent geometry originates in the field of quantum gravity, but this field is also very significant for complex systems such as brain networks. Emergent simplicial geometry refers to the ability of non-equilibrium or equilibrium models to generate simplicial complexes with notable geometric properties by using purely combinatorial rules that make no explicit reference to the network geometry. For instance emergent geometry models should be independent of any possible simplicial complex embedding.

Recently a series of works [29–31] has proposed a theoretical framework called Network Geometry with Flavor that captures the fundamental mechanism of emergent hyperbolic geometry. This framework opens a new perspective into the long-standing problem of emergent geometry and has possible implications ranging from quantum gravity to complex systems. Additionally, this framework generates simplicial networks whose underlying network structure displays all the statistical properties of complex networks including scale-free degree distribution, high clustering coefficient, small-world diameter and significant community structure. The resulting simplicial complexes can reveal distinct geometrical features including a finite spectral dimension [32, 33]. The spectral dimension [34] characterizes the slow relaxation of diffusion processes to their equilibrium steady-state distribution, similarly to what happens for finite-dimensional Euclidean networks. However, higher-order networks with finite spectral dimensions might dramatically differ from Euclidean networks. In fact a finite spectral dimension can co-exist with small-world properties (including an infinite Hausdorff dimension) and a non-trivial community structure. The intrinsic geometrical nature of simplicial complexes with finite spectral dimensions can have a profound effect on dynamical processes such as diffusion and synchronization [35, 36]. In particular, if the spectral dimension d_S is smaller than four, $d_S \leq 4$, it is not possible to observe a synchronized phase of the Kuramoto dynamics and strong spatio-temporal fluctuations are observed instead.

Hyperbolic simplicial geometry also has an important effect on percolation processes. Indeed, percolation on hyperbolic simplicial complexes can display more than one transition and critical behavior at the emergence of the extensive component that deviates from the standard second-order continuous transition. Indeed discontinuous transitions or continuouss transitions with non-trivial critical behavior can be found, depending on the geometry of the higher-order network [37].

1.5 The Advantages of Using Simplicial Complexes and the Outline of the Element Structure

Beside allowing a full topological analysis of higher-order networks, simplicial complexes have the following two advantages: they capture the many-body interactions of a complex system and they allow us to uncover the important role that simplicial topology and simplicial geometry have in dynamics. So far our understanding of the interaction between structure and dynamics has focused on the combinatorial properties of networks (such as their degree distribution) and some of their spectral properties [3, 5]. Study of the interplay between higher-order networks starts to reveal a much richer picture summarized in the diagram presented in Figure 1 in which discrete simplicial network geometry and topology provides new clues to interpret higher-order dynamics. This very innovative framework is emerging from recent research on higher-order networks and has the potential to significantly change the way in which

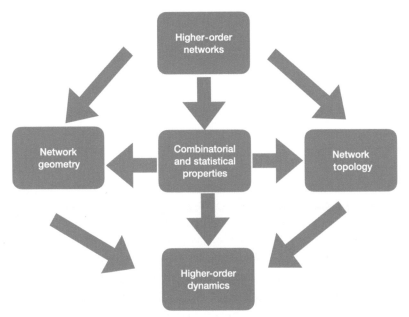

Figure 1 The interplay between higher-order structure and dynamics is mediated by the higher-order combinatorial and statistical properties combined with simplicial network topology and geometry.

we investigate the interplay between structure and dynamics in complex systems. In this Element our goal is to provide the fundamental tools to understand the current research in the field and to make the next steps in this wonderful world of higher-order networks. The Element will introduce important aspects of discrete topology and discrete geometry in a pedagogical way accessible to the interdisciplinary audience of PhD students and researchers in network science.

The Element is structured as follows: in Section 2 we will provide the mathematical definitions of simplicial complexes and discuss their combinatorial and statistical properties, covering generalized degrees and the maximum entropy models of simplicial complexes; Section 3 will cover the basic elements of simplicial network topology, ranging from Topological Data Analysis (TDA) of simplicial complexes, to properties of the higher-order Laplacians; Section 4 is devoted to simplicial network geometry; Section 5 discusses models of emergent geometry including Network Geometry with Flavor; Sections 6, 7, 8 discuss higher-order dynamics including synchronization, percolation and contagion models; finally in Section 9 we provide concluding remarks. The Appendices provide further useful details on the material presented in the main body of this work.

Due to space limitations we have adopted a style that favors coherence of narrative over providing an exhaustive review of all the papers on the subject. Therefore we regret that we have not been able to cover all the growing literature on the subject.

2 Combinatorial and Statistical Properties of Simplicial Complexes

2.1 Mathematical Definitions

2.1.1 Basic Properties of Simplicial Complexes and Hypergraphs

A network is a graph $G = (V, E)$ formed by a set of nodes V and a set of links E that represent the elements of a complex system and their interactions, respectively. Networks are ubiquitous and include systems as different as the WWW (web graphs), infrastructures (such as airport networks or road networks) and biological networks (such as the brain or the protein interaction network in the cell). Networks are pivotal to capturing the architecture of complex systems; however, they have the important limitation that they cannot be used to capture the higher-order interactions. In order to encode for the many-body interactions between the elements of a complex system, higher-order networks need to be used. A powerful mathematical framework to describe higher-order networks is provided by simplicial complexes. Simplicial complexes are formed by a set of simplices. The simplices indicate the interactions existing between two or more nodes and are defined as follows:

SIMPLICES

A d-dimensional simplex α (also indicated as a d-simplex α) is formed by a set of $(d + 1)$ interacting nodes

$$\alpha = [v_0, v_1, v_2 \ldots, v_d].$$

It describes a many-body interaction between the nodes.
It allows for a topological and a geometrical interpretation of the simplex.

For instance, a node is a 0-simplex, a link is a 1-simplex, a triangle is a 2-simplex a tetrahedron is a 3-simplex and so on (see Figure 2).

FACES

A face of a d-dimensional simplex α is a simplex α' formed by a proper subset of nodes of the simplex, i.e. $\alpha' \subset \alpha$.

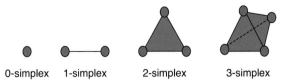

0-simplex   1-simplex   2-simplex   3-simplex

Figure 2 A 0-simplex is a node, a 1-simplex is a link, a 2-simplex is a triangle, a 3-simplex is a tethrahedron and so on.

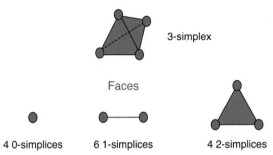

3-simplex

Faces

4 0-simplices   6 1-simplices   4 2-simplices

Figure 3 The faces of a 3-simplex (tetrahedron) are four 0-simplices (nodes), six links (1-simplices) and four triangles (2-simplices).

For instance the faces of a 2-simplex $[v_0, v_1, v_2]$ include three nodes $[v_0]$, $[v_1]$, $[v_2]$ and three links $[v_0, v_1], [v_0, v_2], [v_1, v_2]$. Similarly, in Figure 3 we characterize the faces of a tetrahedron.

The simplices constitute the building blocks of simplicial complexes.

SIMPLICIAL COMPLEX

A simplicial complex $\mathcal{K}$ is formed by a set of simplices that is closed under the inclusion of the faces of each simplex.

The dimension d of a simplicial complex is the largest dimension of its simplices.

Simplicial complexes represent higher-order networks, which include interactions between two or more nodes, described by simplices. In more stringent mathematical terms a simplicial complex $\mathcal{K}$ is a a set of simplices that satisfy the following two conditions:

(a) if a simplex α belongs to the simplicial complex, i.e. $\alpha \in \mathcal{K}$, then any face α' of the simplex α is also included in the simplicial complex, i.e. if $\alpha' \subset \alpha$ then $\alpha' \in \mathcal{K}$;

(b) given two simplices of the simplicial complex $\alpha \in \mathcal{K}$ and $\alpha' \in \mathcal{K}$ then either their intersection belongs to the simplicial complex, i.e. $\alpha \cap \alpha' \in \mathcal{K}$, or their intersection is null, i.e. $\alpha \cap \alpha' = \emptyset$.

Here and in the future we will indicate with N the total number of nodes in the simplicial complex and we will indicate with $N_{[m]}$ the total number of m-dimensional simplices in the simplicial complex (note that $N_{[0]} = N$). Furthermore we will indicate with $\mathcal{Q}_m(N)$ the set of all possible and distinct m-dimensional simplices that can be present in a simplicial complex $\mathcal{K}$ including N nodes. With $\mathcal{S}_m(\mathcal{K})$ we will indicate instead the set of all m-dimensional simplices present in $\mathcal{K}$.

Among the simplices of a simplicial complex, the facets play a very relevant role.

FACET

A facet is a simplex of a simplicial complex that is not a face of any other simplex. Therefore a simplicial complex is fully determined by the sequence of its facets.

A very interesting class of simplicial complexes are pure simplicial complexes.

PURE SIMPLICIAL COMPLEXES

A pure d-dimensional simplicial complex is formed by a set of d-dimensional simplices and their faces.
Therefore pure d-dimensional simplicial complexes admit as facets only d-dimensional simplices.

This implies that pure d-dimensional simplicial complexes are formed exclusively by gluing d-dimensional simplices along their faces. In Figure 4 we show an example of simplicial complex that is pure and an example of a simplicial complex that it is not pure.

An interesting question is whether it is possible to convert a simplicial complex into a network and vice versa and how much information is lost/retained

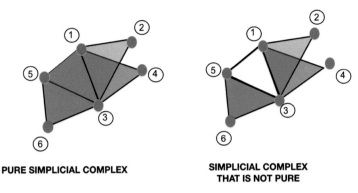

PURE SIMPLICIAL COMPLEX

**SIMPLICIAL COMPLEX
THAT IS NOT PURE**

Figure 4 An example of a 2-dimensional simplicial complex that is pure and an example of a 2-dimensional simplicial complex that is not pure.

in the process. Given a simplicial complex it is always possible to extract a network known as the 1-*skeleton* of the simplicial complex by considering exclusively the nodes and links belonging to the simplicial complex. Conversely, given a network, it is possible to derive deterministically a simplicial complex called the *clique complex* of the network. The *clique complex* is obtained from a network by taking every $(d+1)$-clique in a simplex of dimension d. The clique complex is a simplicial complex. In fact, if a simplex is included in a clique complex, then all its subsimplices are also included. Moreover any two simplices of the clique complex have an intersection that is either the null set or a simplex of the clique complex.

Hypergraphs are alternative representations of higher-order networks that can be used instead of simplicial complexes.

HYPERGRAPH

A hypergraph $\mathcal{G} = (V, E_H)$ is defined by a set V of N nodes and a set E_H of hyperedges, where an $(m+1)$-hyperedge indicates a set of $m+1$ nodes

$$e = [v_0, v_1, v_2, \ldots, v_m],$$

with generic values of $1 \leq m < N$.
A hyperedge describes the many-body interaction between the nodes.

As mathematical objects simplicial complexes are distinct from hypergraphs, the difference being that simplicial complexes include all the subsets of a given simplex. From a network science perspective a given dataset including higher-order interactions can be described either as a simplicial complex or as a hypergraph. However, it might be argued that in a simplicial complex description of a higher-order network dataset we can lose some information. For instance a collaboration network is a good example of a hypergraph where hyperedges correspond to the fact that the considered set of authors (nodes) have published at least a paper together. In this context having a hyperedge connecting three authors indicates that the three authors have co-authored at least a paper together. However, the existence of this three-body interaction does not imply that each scientist has also co-authored a two-author paper with each other scientist in the trio. Therefore by using simplicial complexes to model a collaboration network, we essentially retain only information about the facets of the collaboration while losing detailed information about which lower-dimensional simplex actually indicates a real collaboration. On the other side simplicial complexes allow the use of the very powerful tools of simplicial network topology and geometry that can be a game changer, revealing a much richer interplay between network topology and geometry.

We really believe that it is important for the network scientist to use the right tool for the right problem. It turns out that in many contexts relevant to the network scientist the alternative representations (hypergraph and simplicial complexes) are actually interchangeable. In the cases where they are not, it is important to be able to understand the benefits of one or the other representation and how the results obtained in one framework can be relevant to informing the investigation using the alternative representation.

In this Element, as we want to introduce the reader to the wonderful world of simplicial network topology and geometry, we will mostly cover simplicial complexes; however, results that apply to hypergraphs will also be covered.

2.1.2 Cell Complexes

Cell complexes (or CW complexes where C stands for *closure finite* and W stands for *weak topology*) are a generalization of simplicial complexes that are not exclusively formed by simplices but instead can be formed by basic building blocks called *cells*. Cells describe many-body interactions that are weaker than those of simplicial complexes, and they have a 1-skeleton that differs from a clique. This means that a square can be interpreted as a cell of four-body interactions whose faces are just four links. This can be useful in some situations such as social interaction networks where, for instance, a discussion group can be formed by four people not all having a pairwise social tie with everybody else in the group, or in protein interactions networks where not all the proteins of a protein complex bind pairwise to each other.

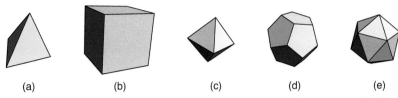

(a)   (b)   (c)   (d)   (e)

Figure 5 All the regular polytopes in $d = 3$ (Platonic solids): (a) tetrahedron, (b) cube, (c) octahedron, (d) dodecahedron, (e) icosahedron.

Source: Reprinted from [32].

Mathematically, a d-dimensional cell is a d-dimensional convex polytope and in general open d-dimensional cells are topological spaces homeomorphic to an open ball. Therefore 0-dimensional cells are nodes and 1-dimensional cells are links, and therefore do not differ from 0-dimensional and 1-dimensional simplices. However, 2-dimensional cells include m-polygons such as triangles (2-dimensional simplices), squares, pentagons etc. Similarly, 3-dimensional cells include the Platonic solids, such as tetrahedra (3-dimensional simplices), cubes, octahedra, dodecahedra and icosahedra (see Figure 5). Interestingly, in dimension $d = 4$ there are more regular polytopes than in dimension $d = 3$ (being 6), but for any dimension $d > 4$ there are only three types of regular (convex) polytopes: the simplex, the hypercube and the orthoplex.

A cell complex $\hat{\mathcal{K}}$ has the following two properties:

(a) it is formed by a set of cells that is closure-finite, meaning that every cell is covered by a finite union of open cells;

(b) given two cells of the cell complex $\alpha \in \hat{\mathcal{K}}$ and $\alpha' \in \hat{\mathcal{K}}$ then either their intersection belongs to the cell complex, i.e. $\alpha \cap \alpha' \in \hat{\mathcal{K}}$, or their intersection is a null set, i.e. $\alpha \cap \alpha' = \emptyset$.

In this Element we will discuss mostly the properties of simplicial complexes; however, in a number of places we will refer to results applying to more general cell complexes.

2.2 Generalized Degrees of Simplicial Complexes

A key local structural property of networks is the degree of the nodes. The degree of a node characterizes only the local structure of the network around the node – its number of interactions. However, the statistical properties associated with the degree are important global properties of the network that can significantly affect its global dynamics, as in the case of scale-free degree distributions [1]. It is therefore natural to desire to extend the notion

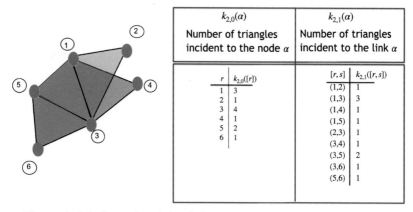

$k_{2,0}(\alpha)$	$k_{2,1}(\alpha)$
Number of triangles incident to the node α	Number of triangles incident to the link α

r	$k_{2,0}([r])$		$[r,s]$	$k_{2,1}([r,s])$
1	3		(1,2)	1
2	1		(1,3)	3
3	4		(1,4)	1
4	1		(1,5)	1
5	2		(2,3)	1
6	1		(3,4)	1
			(3,5)	2
			(3,6)	1
			(5,6)	1

Figure 6 A 2-dimensional simplicial complex is shown together with its generalized degree sequences $k_{2,0}$ and $k_{2,1}$ of the nodes and links, respectively.

of degrees to simplicial complexes. The *generalized degrees* [12, 29, 39] are the fundamental combinatorial properties describing the structure of simplicial complexes. Interestingly, in a simplicial complex not only nodes, but also links and higher-dimensional simplices can be associated with generalized degrees.

> GENERALIZED DEGREES AND FACET SIZES
> The generalized degree [12, 29, 39] $k_{d,m}(\alpha)$ of an m-dimensional simplex α indicates the number of d-dimensional simplices incident to the m-simplex α.

We note that $k_{1,0}([r])$ reduces to the degree of node r in the 1-skeleton of the simplicial complex. In Figure 6 we show a 2-dimensional simplicial complex together with the list of the generalized degrees $k_{2,1}$ and $k_{2,0}$ of its nodes and links, respectively. Note that while the generalized degrees $k_{d,m}$ can be defined for every pair of dimensions (d, m) with $d \neq m$, the generalized degrees with $d < m$ are trivial and do not depend on the simplicial complex structure, indeed they characterize how many d-dimensional faces an m-dimensional simplex has. Therefore, for $d < m$ we have

$$k_{d,m}(\alpha) = \binom{m+1}{d+1},$$

for every m-dimensional simplex of any simplicial complex. It follows that the relevant generalized degrees $k_{d,m}(\alpha)$ for describing the different structures of simplicial complexes have $d > m$. The generalized degrees are of fundamental importance for capturing the combinatorial and statistical properties of simplicial complex data and of models of simplicial complexes. Interestingly, a simplicial complex can display very different generalized degree distributions $k_{d,m}$ if one considers different dimensions m, as we will demonstrate in the case of the model Network Geometry with Flavor (see Section 5.3). The generalized degrees can be also used to determine the combinatorial conditions for observing a discrete manifold (see Section 4.1). They are therefore an important combinatorial aspect of simplicial complexes that is related to their discrete network geometry.

In addition to generalized degrees, another useful statistical property of simplicial complexes is the distribution of the size of their facets. For example, in the collaboration network this captures the distribution of the largest collaborations involving a set of authors [40].

2.3 Pure Simplicial Complexes and Their Tensorial Representation

Pure d-dimensional simplicial complexes have a structure that can be fully captured by a $(d+1)$-dimensional tensor $\mathbf{a}^{[d]}$ called an *adjacency tensor*.

The adjacency tensor $\mathbf{a}^{[d]}$ has elements $a_\alpha^{[d]} \in \{0,1\}$ indicating, for each possible d-dimensional simplex $\alpha \in \mathcal{Q}_d(N)$, if the simplex is present ($a_\alpha = 1$) or absent ($a_\alpha = 0$) in the simplicial complex $\mathcal{K}$, i.e.

$$a_\alpha^{[d]} = \begin{cases} 1 & \text{if } \alpha \in \mathcal{S}_d(\mathcal{K}), \\ 0 & \text{otherwise.} \end{cases} \tag{2.1}$$

The adjacency matrix tensor is symmetric under the permutation of the order of the nodes in the simplices, for instance, if in a 2-dimensional simplical complex we have $a_{rsq}^{[2]} = a_{sqr}^{[2]} = a_{srq}^{[2]}$ and so on for each possible permutation of the three indices forming the simplex $\alpha = [r, s, q]$.

2.4 Generalized Degrees of Pure Simplicial Complexes

2.4.1 General Properties

The generalized degrees $k_{d,m}(\alpha)$ of a pure d-dimensional simplicial complex can be defined in terms of the adjacency tensor $\mathbf{a}^{[d]}$ as

$$k_{d,m}(\alpha) = \sum_{\alpha' \in \mathcal{Q}_d(N) | \alpha' \supseteq \alpha} a_{\alpha'}^{[d]}. \tag{2.2}$$

The generalized degrees obey a nice combinatorial relation as they are not independent of each other. In fact, the generalized degree of an m-face α is related to the generalized degree of the m'-dimensional faces incident to it, with $m' > m$, by the simple combinatorial relation

$$k_{d,m}(\alpha) = \frac{1}{\binom{d-m}{m'-m}} \sum_{\alpha' \in \mathcal{Q}_d(N)|\alpha' \supseteq \alpha} k_{d,m'}(\alpha'). \qquad (2.3)$$

Moreover, since every d-dimensional simplex belongs to $\binom{d+1}{m+1}$ m-dimensional faces, in a simplicial complex with M d-dimensional simplices we have

$$\sum_{\alpha \in \mathcal{S}_m(\mathcal{K})} k_{d,m}(\alpha) = \binom{d+1}{m+1} M. \qquad (2.4)$$

2.4.2 Case of a Simplicial Complex of Dimension $d = 1$

Simplicial complexes of dimension $d = 1$ are networks and therefore are formed exclusively by nodes and links. The adjacency tensor of the 1-dimensional simplicial complex reduces to the adjacency matrix $\mathbf{a}^{[1]}$ (also indicated as $\mathbf{A}$), whose elements $a_{rs}^{[1]}$ (or A_{rs}) indicate whether the link $[r, s]$ is present or not in the network. In this case, the generalized degree $k_{1,0}([r])$ of a node r simply indicates its degree (also indicated as k_r), i.e. the number of links incident to it,

$$k_{1,0}([r]) = \sum_{s=1}^{N} a_{rs}^{[1]}. \qquad (2.5)$$

2.4.3 Case of a Simplicial Complex of Dimension $d = 2$

Pure simplicial complexes of dimension $d = 2$ are formed exclusively by a set of triangles and their faces (nodes and links). A 2-dimensional pure simplicial complex is determined by the adjacency tensor $\mathbf{a}^{[2]}$ of elements $a_{rsq}^{[2]} = 1$ if the triangle $\alpha = [r, s, q]$ belongs to the simplicial complex and $a_{rsq}^{[2]} = 0$ if it does not belong to the simplicial complex. The generalized degree $k_{2,0}([r])$ of node r is given by

$$k_{2,0}([r]) = \sum_{s<q} a_{rsq}^{[2]}, \qquad (2.6)$$

while the generalized degree $k_{2,1}([r, s])$ of a link $\alpha = [r, s]$ is given by

$$k_{2,1}([r, s]) = \sum_{q=1}^{N} a_{rsq}^{[2]}. \qquad (2.7)$$

The generalized degree $k_{2,0}([r])$ of node r indicates the number of triangles incident to it, while the generalized degree $k_{2,1}([r,s])$ of the link $[r,s]$ indicates the number of triangles incident to the link. The generalized degree of the nodes is related to the generalized degree of the links. In fact it is easy to see that

$$k_{2,0}([r]) = \sum_{s<q} a_{rsq}^{[2]} = \frac{1}{2}\sum_{s,q} a_{rsq}^{[2]} = \frac{1}{2}\sum_{s=1}^{N} k_{2,1}([r,s]). \tag{2.8}$$

Since each triangle is incident to three nodes, we have

$$\sum_{r=1}^{N} k_{2,0}([r]) = 3M, \tag{2.9}$$

where M is the number of 2-dimensional simplices in the simplicial complex.

2.5 Clique Complexes of Random Uncorrelated Networks

Clique complexes are a simple way to generate simplicial complexes starting from network data. As network data is currently more abundant than higher-order network data, it is important to investigate here the properties of the clique complexes of the most simple models of networks, i.e. random uncorrelated networks. Random uncorrelated networks are sparse networks, indicating that the number of links scales as the number of nodes in the network. Moreover, random uncorrelated networks are random networks with a given degree sequence, as long at the degrees of the network have a structural cutoff. The structural cutoff imposes that all the degrees of the network should be much lower than the structural cutoff, i.e.

$$k_r \ll K = \sqrt{\langle k \rangle N}, \tag{2.10}$$

for all nodes $r \in \{1, 2, \ldots\}$, where here $\langle k \rangle$ indicates the average degree of the network. In this limit the maximum entropy network ensemble enforcing the given degree sequence describes a null model in which there are no degree–degree correlations, i.e. the degree of the node at one end of a link is not correlated with the degree of the node at the other end of the link.

If one performs the clique complex of this ensemble, every $(d+1)$-clique is reduced to a d-dimensional simplex of the clique complex. For Erdös–Rényi graphs in which the number of links L scales as the number of nodes N (Poisson networks), the expected number of cliques of size greater than 3 is null in the large N limit, implying that the clique complex is at most 2-dimensional [41]. However, the situation changes significantly if one considers an uncorrelated network with power-law degree distribution $P(k) = Ck^{-\gamma}$ and exponent

$\gamma \in (2,3)$ [42]. Indeed, in this case the expected number $\mathcal{N}_d$ of d-simplices in the clique complex of the network scales, as long as $d > 1$, as

$$\mathcal{N}_d = \mathcal{O}(N^{\xi}) \tag{2.11}$$

with

$$\xi = \frac{1}{2}(d+1)(d+1-\gamma), \tag{2.12}$$

so it is diverging in the large network limit $N \to \infty$. For $\gamma = 3$, a logarithmic scaling is observed:

$$\mathcal{N}_d = \mathcal{O}((\ln N)^{d+1}). \tag{2.13}$$

Moreover, and even more surprisingly, the clique number of the random scale-free network diverges [42], i.e. the size of the largest clique of the network skeleton of the simplicial complex diverges with $N \to \infty$ for $\gamma \in (2,3]$. This implies that in a clique complex of a random uncorrelated scale-free network, the dimension of the largest facet diverges. This phenomenon is due to the fact that uncorrelated scale-free networks display a hierarchical structure (also revealed by their core structure) and it becomes easier for nodes of high degrees to form cliques.

It follows that if we consider the clique complex of a network with broad degree distribution, and also if the network is sparse, we expect that the dimension of the obtained simplicial complex is significantly large. Therefore the clique complex of a network with broad degree distribution is expected to have very significant differences with respect to the clique complex of a Poisson network with the same average degree.

2.6 Maximum Entropy Ensembles of Simplicial Complex

2.6.1 Microcanonical and Canonical Ensembles of Simplicial Complexes

A combination of combinatorial and statistical arguments are at the very heart of information theory of simplicial complexes. This theory has the aim to provide null models of simplicial complexes that are the least biased given a set of constraints [12]. These models are obtained using the Maximum Entropy Principle, which is one of the pillars of information theory [6, 43, 44]. First of all we consider an ensemble of simplicial complexes, formed by assigning a probability $P(\mathcal{K})$ to every simplicial complex $\mathcal{K}$ belonging to the set $\Omega_{\mathcal{K}}$ of all possible simplicial complexes. The entropy of the ensemble of simplicial complexes is given by

$$S = -\sum_{\mathcal{K} \in \Omega_{\mathcal{K}}} P(\mathcal{K}) \ln P(\mathcal{K}), \qquad (2.14)$$

where we use the standard convention that if $P(\mathcal{K}) = 0$, then we take $P(\mathcal{K}) \ln P(\mathcal{K}) = 0 \ln 0 = 0$. The entropy of the ensemble of simplicial complexes characterizes the logarithm of the typical number of simplices in the ensemble.

According to the Maximum Entropy Principle, the least-biased ensemble is one satisfying a given set of constraints, characterized by a probability $P(\mathcal{K})$ that can be obtained by maximizing the entropy S, given the constraints. The entropy of maximum-entropy simplicial complexes is a fundamental information theory measure that can be used to quantify the information content of the constraints. In fact constraints that carry a high information content are more difficult to satisfy by a random simplicial complex and correspond to a smaller entropy of the ensemble. On the other hand, constraints that do not limit the structure of simplicial complexes much are more easy to satisfy and correspond to a higher value of entropy.

Recently [12, 45], it has been shown that maximum-entropy models of networks and generalized network structures can be constructed using an analogy with statistical mechanics. In statistical mechanics [46] a distinction is made between canonical and microcanonical ensembles, which consider configurations of a dynamical system that are compatible with a given value of the energy and with a given expected value of the energy. Similarly, when constructing ensembles of networks or of simplicial complexes, we can consider ensembles with a given set of hard constraints (satisfied in any single instance) or a given set of soft constraints (satisfied on average over all the instances of the ensemble).

For simplicial complex ensembles we can consider a given set of observables $F_\mu(\mathcal{K})$ with $\mu = 1, 2, \dots \hat{P}$, which might indicate, for instance, the total number of d-dimensional simplices or the generalized degree of a node in a pure d-dimensional simplicial complex. From these observables it is possible to consider a set of hard constraints given by

$$F_\mu(\mathcal{K}) = C_\mu, \qquad (2.15)$$

which needs to be satisfied for every simplicial complex $\mathcal{K}$ represented in the ensemble, i.e. for every simplicial complex $\mathcal{K}$ having non-zero probability in the ensemble. Alternatively, it is possible to consider a set of soft constraints that enforce limits on average on the ensemble, i.e.

$$\sum_{\Omega_{\mathcal{K}}} P(\mathcal{K}) F_\mu(\mathcal{K}) = \bar{C}_\mu. \qquad (2.16)$$

Hard and soft constraints corresponding to the same choice of observables $F_\mu(\mathcal{K})$ and having $C_\mu = \bar{C}_\mu$ are called *conjugated constraints*. The maximum entropy ensembles satisfying the soft constraints are called *canonical ensembles* and are characterized by a probability $P_C(\mathcal{K})$ that follows a Gibbs distribution and is therefore given by (see Appendix A for details of the derivation)

$$P_C(\mathcal{K}) = \frac{1}{Z_C} e^{-\sum_{\mu=1}^{\hat{P}} \lambda_\mu F_\mu(\mathcal{K})}, \tag{2.17}$$

where the parameters λ_μ are the Lagrangian multipliers enforcing the constraints and are fixed by imposing Eq. (2.16), while Z_C is a normalization constant also called the *partition function*. The maximum-entropy ensembles satisfying the hard constraints are called *microcanonical ensembles* and are characterized by a probability $P_M(\mathcal{K})$ that is uniform over all the simplicial complexes satisfying the hard constraints (see Appendix A for details of the derivation), i.e.

$$P_M(\mathcal{K}) = \frac{1}{Z_M} \prod_{\mu=1}^{\hat{P}} \delta(C_\mu, F_\mu(\mathcal{K})), \tag{2.18}$$

where here and in the following $\delta(x, y)$ indicates the Kronecker delta, with $\delta(x, y) = 1$ if $x = y$ and otherwise $\delta(x, y) = 0$, and where Z_M indicates the number $\mathcal{N}$ of simplicial complexes satisfying the hard constraints, i.e. $Z_M = \mathcal{N}$. The entropy Σ of the microcanonical ensemble is therefore given by the logarithm of the number of simplicial complexes in the ensemble,

$$\Sigma = \ln \mathcal{N}. \tag{2.19}$$

Canonical and microcanonical ensembles satisfying conjugated constraints are *conjugated ensembles*. If we indicate with S the entropy of the canonical ensemble and with Σ the entropy of the conjugated microcanonical ensemble we have [12, 47]

$$\Sigma = S - \hat{\Omega}, \tag{2.20}$$

where $\hat{\Omega}$ is defined (see Appedix A for details of the derivation) as

$$\hat{\Omega} = -\ln \left[\sum_{\mathcal{K} \in \Omega_{\mathcal{K}}} P_C(\mathcal{K}) \prod_{\mu=1}^{\hat{P}} \delta(C_\mu, F_\mu(\mathcal{K})) \right]. \tag{2.21}$$

Therefore $\hat{\Omega}$ is a measure of how much the canonical ensemble deviates from the microcanonical ensemble. In fact, $\hat{\Omega}$ gives the absolute value of the logarithm of the probability that in the canonical ensemble the hard constraints are satisfied. If $\hat{\Omega}$ is subextensive, i.e. $\hat{\Omega} = o(N)$, then the conjugated ensembles are statistically equivalent, meaning that they have the same statistical properties in the limit of large simplicial complexes, i.e. $N \to \infty$. Conversely, if $\hat{\Omega}$ is extensive, i.e. $\hat{\Omega} = \mathcal{O}(N)$ then the two ensembles are not statistically equivalent [12, 47]. In simplicial complexes, as in networks, this latter scenario occurs in cases in which we fix an extensive number of constraints, i.e. a number of constraints $\hat{P}$ that is proportional to N [12]. This includes, for instance, the case in which we impose the expected generalized degree sequence of the nodes.

2.6.2 Canonical Ensemble of Simplicial Complexes with Given Generalized Degree Sequence of the Nodes

Maximum-entropy simplicial complexes can be used to define simplicial complex models that are the higher-order equivalent of Erdös and Rényi random graphs [48, 49]. Here we focus on ensembles that capture the heterogeneities present in simplicial complexes and that are most useful to the network scientist. Specifically we consider canonical ensembles of pure d-dimensional simplicial complexes with given expected generalized degree sequences of the nodes $\{\bar{k}_{d,0}([r])\}$ first studied in Ref. [12]. We identify every pure (labelled) simplicial complex with its adjacency tensor $\mathbf{a}^{[d]}$ indicated simply as $\mathbf{a}$ for a more concise notation. The maximum entropy ensemble of simplicial complexes is determined when we know the probability $P(\mathcal{K}) = P(\mathbf{a})$ of every simplicial complex $\mathcal{K}$ with adjacency tensor $\mathbf{a}$.

We can find $P(\mathbf{a})$ of the canonical ensembles of pure d-dimensional simplicial complexes with given expected generalized degree sequences of the nodes $\{\bar{k}_{d,0}([r])\}$ by maximizing the entropy S under the constraints

$$\sum_{\mathbf{a}} P(\mathbf{a}) F_r(\mathbf{a}) = \bar{k}_{d,0}([r]), \tag{2.22}$$

where $F_r(\mathbf{a})$ is given by

$$F_r(\mathbf{a}) = \sum_{i_1 < i_2 < \ldots < i_d} a_{r i_1 i_2, i_3 \ldots i_d} \tag{2.23}$$

and where $\bar{k}_{d,0}([r])$ indicates the expected number of d-dimensional simplices incident to each node $r \in \{1, 2, \ldots, N\}$.

We will consider here exclusively the sparse regime, which is relevant for most of the applications of complex networks, in which the number of simplices M is of the same order of magnitude as the number of nodes $M \propto N$. In this limit the average expected generalized degree $\langle \bar{k}_{d,0}([r]) \rangle$ is independent of the network size. The maximum entropy distribution $P(\mathbf{a})$ obeys the Gibbs measure given by Eq. (2.17) corresponding to the choice of constraints given by Eqs. (2.22)–(2.23). The marginal probability $p_\alpha = \sum_{\mathcal{K} \in \Omega_\mathcal{K}} a_\alpha P(\mathcal{K})$ of a

d-dimensional simplex α indicates the probability that the simplex belongs to the simplicial complex and is given by

$$p_\alpha = \frac{e^{-\sum_{r \subset \alpha} \lambda_r}}{1 + e^{-\sum_{r \subset \alpha} \lambda_r}}. \tag{2.24}$$

Note that the constraints of the ensemble can be expressed simply in terms of these marginals, implying that the equations that the Lagrangian multipliers $\{\lambda_r\}$ need to satisfy are

$$\bar{k}_{d,0}([r]) = \sum_{\alpha | r \subset \alpha} p_\alpha = \sum_{\alpha | r \subset \alpha} \frac{e^{-\sum_{s \subset \alpha} \lambda_s}}{1 + e^{-\sum_{s \subset \alpha} \lambda_s}}. \tag{2.25}$$

The expression (Eq.(2.24)) for the marginal probability p_α reduces for $d = 1$ to the known marginal probability p_{rs} of a link $\alpha = [r, s]$ in the canonical network ensemble

$$p_{rs} = \frac{e^{-\lambda_r - \lambda_s}}{1 + e^{-\lambda_r - \lambda_s}}. \tag{2.26}$$

For $d = 2$, instead the marginal probability p_{rsq} of a triangle $\alpha = [r, s, q]$ reads

$$p_{rsq} = \frac{e^{-\lambda_r - \lambda_s - \lambda_q}}{1 + e^{-\lambda_r - \lambda_s - \lambda_q}}. \tag{2.27}$$

The fact that in general these marginal probabilities do not factorize into factors depending exclusively on a single node of the simplex indicates that the ensemble of simplicial complexes displays natural generalized degree–degree correlations. These natural correlations are induced among the generalized degrees of the nodes connected by a simplex and they are the higher-order counterpart of the natural correlations that are well known to exist in maximum-entropy ensembles of networks with a given expected degree distribution. A study of generalized degree–degree correlations in the canonical ensemble of simplicial complexes (see Ref. [12]) reveals that the simplicial correlations found in these ensembles are more pronounced than in the ensemble of networks (simplicial complexes of dimension $d = 1$).

In the presence of the structural cutoff, K, i.e., when the generalized degree of the nodes obey

$$\bar{k}_{d,0}([r]) \ll K = \left(\frac{\left(\langle \bar{k}_{d,0}([r]) \rangle N \right)^d}{d!} \right)^{1/(d+1)}, \tag{2.28}$$

the marginal p_α factorizes and becomes proportional to the product of the expected generalized degree of its nodes [12], i.e.

$$p_\alpha = d! \frac{\prod_{r\subset\alpha} \bar{k}_{d,0}([r])}{\left(\langle \bar{k}_{d,0}([r])\rangle N\right)^d}. \tag{2.29}$$

This implies that in this limit the generalized degree–degree correlations vanish and we can consider the canonical ensemble of simplicial complexes null models of uncorrelated simplicial complexes. This expression for the marginal probability is very useful for studying dynamical processes on simplicial complexes. For $d = 1$ the expression of the uncorrelated marginal probability p_{rs} of a link $\alpha = [r,s]$ reduces to the well-known expression

$$p_{rs} = \frac{\bar{k}_{1,0}([r])\bar{k}_{1,0}([s])}{\left(\langle \bar{k}_{1,0}([r])\rangle N\right)}. \tag{2.30}$$

However, for $d = 2$ the uncorrelated marginal probability p_{rsq} of a triangle $\alpha = [r,s,q]$ reads

$$p_{rsq} = 2\frac{\bar{k}_{2,0}([r])\bar{k}_{2,0}([s])\bar{k}_{2,0}([q])}{\left(\langle \bar{k}_{2,0}([r])\rangle N\right)^2}. \tag{2.31}$$

The entropy of the canonical ensemble of simplicial complexes has a very simple expression due to the fact that the constraint enforcing given generalized degrees of the nodes is linear in the adjacency tensor. Indeed we have that the entropy S of the ensemble is fully determined by the marginals p_α and is given by

$$S = - \sum_{\alpha\in Q_d(N)} [p_\alpha \ln p_\alpha + (1-p_\alpha)\ln(1-p_\alpha)]. \tag{2.32}$$

Similarly, the probability distribution of the full simplicial complex can be expressed in terms of the marginal probabilities as

$$P(\mathbf{a}) = \prod_{\alpha\in Q_d(N)} p_\alpha^{a_\alpha}(1-p_\alpha)^{1-a_\alpha}. \tag{2.33}$$

The simplicial complexes in the canonical ensemble can be generated very simply according to the following algorithm:

(1) The marginal probabilities p_α are calculated for each possible d-dimensional simplex α of the simplicial complex by solving Eq. (2.25).
(2) Each possible simplex $\alpha \in Q_d(N)$ of a simplicial complex formed by N nodes is included in the simplicial complex, together with all its faces, with probability p_α.

2.6.3 Configuration Model of Simplicial Complexes

The configuration model of simplicial complexes [12] is the maximum-entropy microcanonical ensemble of simplicial complexes having a given sequence of generalized degrees of the nodes $\{k_{d,0}([r])\}_{r=1,2,\ldots,N}$. The probability $P(\mathcal{K})$ is therefore uniform over all simplicial complexes having the same sequence of generalized degrees of the nodes. The partition function Z_M indicates the number of distinct simplicial complexes having the same generalized degree sequence of the nodes.

Having defined the configuration model of simplicial complexes, one important problem is the formulation of an actual algorithm for generating simplicial complexes belonging to this ensemble. This algorithm builds on the representation of the simplicial complex in terms of a factor graph, i.e. a bipartite graph formed by nodes, and factor nodes (representing d-dimensional simplices, for instance triangles). The algorithm, schematically represented in Figure 7, is defined as follows:

- Starting from a set of N nodes, to each node r a set of $k_{d,0}([r])$ stubs is assigned.
- The stubs are matched to auxiliary factor nodes of degree $d + 1$.
- The obtained factor graph is converted into a simplicial complex. Every set of $(d + 1)$ nodes incident to the same factor node of the factor graph corresponds, in the simplicial complex, to a d-dimensional simplex formed by the same set of nodes.

The matching of the stubs needs to be done randomly; however, some matches are not permitted and if they occur the algorithm needs to be rerun from scratch. The illegal matchings (see Figure 8 for a schematic description) are: matches in which the same set of nodes are matched to more than one factor node or matches in which the same factor node is matched to more than one stub incident to the same node. The code for the configuration model of a simplicial complex is available in the repository [50]. From the algorithm generating single instances of the simplicial networks in the configuration model we deduce two main conclusions:

- Given the generalized degree of the nodes there are, in general, multiple ways to realize the simplicial complex.
- The information encoded in the constraints is captured by the entropy Σ of the ensemble, which is given by the logarithm of the number $\mathcal{N}$ of simplicial complexes that realize a given generalized degree of the nodes, i.e. $\Sigma = \ln \mathcal{N}$.

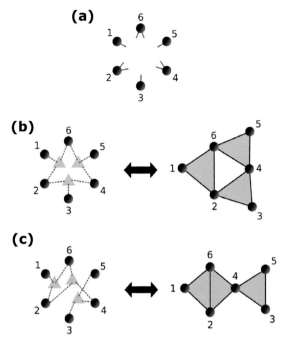

Figure 7 Schematic representation of the algorithm generating single instances of the configuration model of simplicial complexes. Panel (a): each node of the simplicial complex is assigned a number $k_{d,0}(r)$ of stubs. Panels (b) and (c): stubs are matched to factor nodes of degree $d + 1$, and the factor graph is converted into a simplicial complex. It is evident that with this algorithm, different simplicial complexes with the same generalized degree sequence of the nodes can be generated (such as the different simplicial complexes shown in panel (b) and panel (c)).

Source: Reprinted figure with permission from [12] ©Copyright (2016) by the American Physical Society.

The entropy Σ of the configuration model is related to the entropy S of the conjugated canonical ensemble of simplicial complexes with a given expected generalized degree of the nodes by Eq. (2.20), where $\hat{\Omega}$ is given by [12]

$$\hat{\Omega} = -\sum_{r=1}^{N} \ln \left[\frac{(k_{d,0}([r]))^{k_{d,0}([r])}}{k_{d,0}([r])!} e^{-k_{d,0}([r])} \right],$$

as long as the simplicial complex displays the structural cutoff given by Eq. (2.28). From this expression it is possible to deduce the asymptotic number $\mathcal{N}$ of simplicial complexes with a given generalized degree of the nodes, which is given by [12]

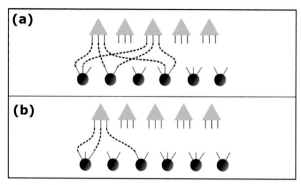

Figure 8 Schematic representation of the illegal matches in the configuration model of a simplicial complex of dimension $d = 2$. Panel (a): It is not possible to match the same set of $(d + 1)$ nodes to more than one factor node. Panel (b): A given factor node cannot be matched to more than one stub of a given node.

Source: Reprinted figure with permission from [12] ©Copyright (2016) by the American Physical Society.

$$\mathcal{N} \sim \frac{[(\langle k_{d,0}\rangle N)!]^{d/(d+1)}}{\prod_{r=0}^{N} k_{d,0}([r])!} \frac{1}{(d!)^{\langle k_{d,0}\rangle N/(d+1)}}$$

$$\times \exp\left[-\frac{d!}{2(d+1)(\langle k_{d,0}\rangle N)^{d-1}}\left(\frac{\langle k_{d,0}^2\rangle}{\langle k_{d,0}\rangle}\right)^{d+1}\right] \qquad (2.34)$$

This formula reveals, among other things, that ensembles of simplicial complexes formed by the same number of d-dimensional simplices but having different generalized degrees of the nodes can have a very different number of instances. Interestingly this intricate combinatorial formula for $\mathcal{N}$ can be explained in terms of the matching algorithm that generates single instances of simplicial complexes in the configuration model (see discussion in [12]). Moreover, this formula in the limit $d = 1$ reduces to the well-known Bender–Canfield formula [51] for the asymptotic number of networks with a given degree sequence, i.e.

$$\mathcal{N} \sim \frac{(\langle k_{1,0}\rangle N)!!}{\prod_{r=0}^{N} k_{1,0}([r])!} \exp\left[-\frac{1}{4}\left(\frac{\langle (k_{1,0})^2\rangle}{\langle k_{1,0}\rangle}\right)^2\right]. \qquad (2.35)$$

2.6.4 From Simplicial Complex Models to the Clique Complex of Their Skeleton

It is interesting to note that the properties of simplicial complexes generated by maximum entropy models might be significantly different from the properties of the clique complexes constructed from their skeleton. We have already discussed in Section 2.5 that the clique complex of a scale-free network (simplicial complex of dimension 1) can have a very large dimensionality; actually it has a dimension that diverges with the network size N. A similar situation can occur if we start from maximum entropy models of simplicial complexes of dimension $d > 1$, we generate their network skeletons (i.e. we retain only their nodes and links) and we then consider the clique complex of this latter structure. In particular, if we start from maximum-entropy models of simplicial complexes with broad or scale-free generalized degrees of the nodes and we follow the procedure described before, we typically end up with simplicial complexes in which the number of d-dimensional simplices strongly deviates from those of the original model. This is a totally consistent result, because simplicial complexes with a broad distribution of the generalized degree of the nodes have a region in which d-dimensional simplices cluster. This is not a shortcoming of the model but might lead to some incongruencies if the modeler is not fully aware of it and desires to move back and forth from the simplicial complex description to the network description.

2.7 From Ensembles of Pure Simplicial Complexes to Ensembles of Hypergraphs

In the previous paragraphs we have discussed how to construct ensembles of pure simplicial complexes with a given generalized degree sequence or expected generalized degree sequence. For $d = 2$ these ensembles have a structure that is very similar to the early model of hypergraphs proposed in Ref. [52]. Most interestingly, the ensembles discussed in the previous section and studied in Ref. [12] are fundamental to generating ensembles of hypergraphs with hyperedges of arbitrary size. Indeed, a maximum-entropy ensemble of hypergraphs with hyperedges of maximum size m can be obtained by aggregating ensembles of pure simplicial complexes of dimension m with $1 \leq m \leq d$ [12]. The construction of a maximum-entropy hypergraph where each node r has $k_{m,0}([r])$ hyperedges of size $m + 1$ can indeed be obtained according to the following algorithm using a multiplex network [6] construction (see Figure 9):

(1) Consider a multiplex network of N nodes and d layers.
(2) Draw each layer m (with $1 \leq m \leq d$) of the multiplex network from a configuration model of pure simplicial complexes of dimension m with given generalized degree sequence of the nodes $\{k_{m,0}([r])\}_{r=1,2,...,N}$.
(3) For each pure m-dimensional simplicial complex, when forming the m-th layer of the multiplex network, retain only the largest facets, (i.e. the simplices of dimension m).

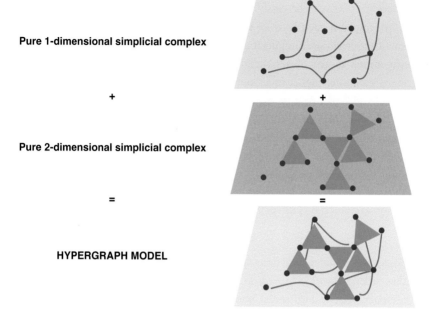

Pure 1-dimensional simplicial complex

+

Pure 2-dimensional simplicial complex

=

HYPERGRAPH MODEL

Figure 9 Multilayer construction of a maximum-entropy hypergraph model in which each node r belongs to $k_{m,0}([r])$ hyperedges of size $m + 1$. In the case shown in the picture $m \in \{1, 2\}$.

(4) Consider the hypergraph aggregating all the layers. In other words, consider the hypergraph whose set of hyperedges is the union of all the facets of each pure m-dimensional simplicial complex of each layer of the multiplex network.

In this ensemble the set of interactions among $m + 1$ nodes is fully captured from the adjacency tensor $\mathbf{a}^{[m]}$. Assuming that the hypergraph is formed by N nodes, each node r (with $r \in \{1, 2 \dots, N\}$) is assigned a vector $\mathbf{k}_r$ of generalized degrees

$$\mathbf{k}_r = (k_{1,0}([r]), k_{2,0}([r]), \dots, k_{d,0}([r])), \tag{2.36}$$

where $k_{m,0}([r])$ indicates the number of $(m + 1)$-hyperedges incident to node r.

When the configuration model of hypergraphs is assumed to be uncorrelated, the probability $p_\alpha^{[m]}$ that a hyperedge $\alpha = [r, v_1, v_2, \dots v_m]$ including node r and m other nodes $\{v_j\}_{j=1\dots,m}$ is present in the hypergraph is given by

$$p_\alpha^{[m]} = \langle a_\alpha^{[m]} \rangle = m! k_{m,0}([r]) \prod_{j=1}^{d} \left(\frac{k_{m,0}([v_j])}{\langle k_{m,0} \rangle N} \right). \tag{2.37}$$

This construction can be easily generalized to generate canonical ensembles of hypergraphs if each layer of the multiplex network construction is drawn from a canonical pure simplicial complex ensemble rather than from the pure configuration model of simplicial complexes. This model and its generalizations capturing higher-order networks formed by collections of arbitrary subgraphs [53] can be very useful to investigate the structure of real higher-order datasets [54] and to study dynamical processes. This ensemble has been recently adopted to study higher-order percolation processes [55]. In Section 8 we will discuss how to treat higher-order contagion models defined on such generalized network structures.

2.8 Temporal Simplicial Complexes

2.8.1 Temporal Social Networks

Face-to-face social interactions are a very important example of temporal higher-order networks. Indeed, in a party, as well as in a coffee break at a conference, face-to-face social interactions typically involve more than just two individuals forming simplices or hyperedges of higher-order networks [56–58]. Interestingly, these higher-order networks are temporal, meaning that higher-order interactions are established in time and have a given duration. The field of temporal networks [59] has revealed many important new phenomena occurring when interactions are dynamically changing in time. A notable property of temporal social networks is their bursty dynamics [56, 60–64] indicating that social interactions are not established at a constant rate in time, and do not have a Poisson duration characterized by a typical time-scale. Rather, social interactions have a duration and an inter-event time distribution which is broad.

Two important questions emerge if we investigate the higher-order structure of these temporal networks: (a) How can we evaluate the predictability of such dynamical networks with information theory tools? (b) What is the dependence of the distribution of duration of higher-order interactions on the number of nodes (individuals) participating in it? These two questions have been addressed in Refs. [57] and [56], whose main results we will discuss in the next two paragraphs.

2.8.2 Entropy of Temporal Simplicial Complexes

In order to quantify the information encoded in the dynamics of higher-order temporal social networks Ref. [57] introduces the notion of entropy of dynamical higher-order social networks. The higher-order temporal social network is defined starting from a static social network G of friendships, collaborations

(a) (b)

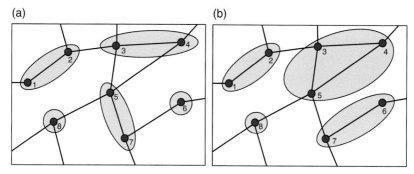

Figure 10 Higher-order temporal social networks are composed of different dynamically changing groups of interacting agents. In panel (a) we allow only for hyperedges of size one or two as typically happens in mobile phone communication. In panel (b) we allow for hyperedges of any size as in face-to-face interactions.

Source: Reprinted from Ref. [32] .

or acquaintances formed by N agents. Any set of agents $v_0, v_1, \ldots v_m$ forming a connected subgraph of the network G is allowed to dynamically interact in a cell of dimension m (a hyperedge of size $m + 1$). Therefore, at any given time the static network G will be partitioned at any point in time, in disconnected hyperedges of interacting agents that form and dissolve in time (for a given snapshot of time see the schematic representation in Figure 10). In order to indicate that a higher-order social interaction is occurring at time t among the agents belonging to the hyperedge $\alpha = [v_0, v_1, \ldots, v_m]$ and that these agents are not interacting with other agents, we write $g_\alpha(t) = 1$, otherwise we put $g_\alpha(t) = 0$. Therefore each agent in a hyperedge of size $m + 1$ is interacting with m other agents, as long as $m > 0$, or non-interacting with any other agent (in the case $m = 0$). It follows that at any given time the tensor g_α satisfies for every node r of the hypergraph

$$\sum_{\{v_j\}\mid[r,v_1,v_2,\ldots,v_m]\in\mathcal{K}} g_{[r,v_1,v_2,\ldots,v_m]}(t) = 1, \tag{2.38}$$

where we indicate with $\mathcal{K}$ the hypergraph formed by any connected subgraph of G. The history $\mathcal{I}_t$ of the dynamical higher-order social network is given by $\mathcal{I}_t = \{g_\alpha(t') \, \forall t' < t\}$. If we indicate by $P(g_\alpha(t) = 1|\mathcal{I}_t)$ the probability that $g_\alpha(t) = 1$ given the history $\mathcal{I}_t$, the likelihood that at time t the social interactions have a hypergraph configuration $g_\alpha(t)$ is given by

$$\mathcal{L} = \prod_{\alpha\in\mathcal{K}} P(g_\alpha(t) = 1|\mathcal{I}_t)^{g_\alpha(t)}. \tag{2.39}$$

The entropy $\hat{S}$ of higher-order temporal social networks characterizes the logarithm of the typical number of hypergraphs that can be expected in the dynamical hypergraph model at time t and is given by $\hat{S} = -\langle \ln \mathcal{L} \rangle_{|\mathcal{I}_t}$, which we can explicitly express as

$$\hat{S} = -\sum_{\alpha \in \mathcal{K}} P(\mathcal{I}_t) P(g_\alpha(t) = 1|\mathcal{I}_t) \ln P(g_\alpha(t) = 1|\mathcal{I}_t). \tag{2.40}$$

According to information theory, if the entropy of higher-order temporal social networks is vanishing, i.e. $\hat{S} = 0$, the hypergraph dynamics are regular and perfectly predictable; if the entropy is larger the number of future possible configurations is growing and the system is less predictable. In Ref. [57] this entropy has been used to characterize the predictability of real higher-order temporal social networks.

2.8.3 Face-to-Face Simplicial Interactions

With recently developed technology it is now possible to obtain detailed data about the statistical properties of higher-order social interactions. In particular, the SocioPattern project [65] has collected data from face-to-face interactions in a number of environments, including conferences, hospitals, museums and so on. This precious data is particularly important to study epidemic spreading in a realistic setting. An important aspect of this data is that it allows testing of a hypothesis on the bursty nature of human face-to-face interactions [56, 62]. The SocioPattern data reveals that pairwise social interactions have durations that are distributed as a power-law [62] and that higher-order social interactions also have a power-law duration [56] with an exponent that is steeper the larger the dimension m of the interacting simplices (the size $m + 1$ of the group of interacting agents). This phenomenology, shown in Figure 11, is captured by the simple model of higher-order temporal social networks proposed in Ref. [56]. The higher volatility of social interactions involving larger simplices is explained by the fact that in larger groups of people, the size of the group changes if any one of its members leaves the group. Therefore, gatherings that involve more individuals have a steeper distribution of contact duration than gatherings involving fewer people.

3 Simplicial Network Topology

3.1 Introduction to Simplicial Network Topology

Topology is the branch of mathematics that is interested in shapes and that characterizes the properties of geometrical objects that are invariant under continuous deformations. This field has the same origin as graph theory, as it is

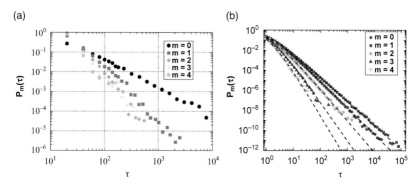

Figure 11 Distribution $P_m(\tau)$ of duration of an m-dimensional simplex representing the temporal social interactions in a SocioPattern experiment (panel a) and in the higher-order temporal network proposed in [56] (panel b). Simplices of higher dimension m have a distribution $P_m(\tau)$ of durations that is steeper than simplices with smaller dimension m.

Source: Reprinted figure with permission from [56] ©Copyright (2011) by the American Physical Society.

traced back to 1756, the year in which Euler published a paper solving the problem of the seven bridges of Könisberg. This famous problem concerns determining a path in the city of Könisberg that starts from one side of the river and goes back to the original point by crossing each of the seven bridges exactly once. It is well known that Euler solved this famous puzzle by mapping it to a graph, disregarding all information about distances between different regions of the city and only capturing their connectivity, i.e. if one region of the city was reachable from another region of the city by crossing a bridge. When topology is used to investigate simplicial and cell complexes, a more rich scenario emerges, where topology can play an even more important role. Interestingly, it was again Euler that first identified the Euler characteristic χ as a topological invariant, noticing that all polyhedra obey the classic polyhedron formula relating the number of vertices V, edges E and faces F by $V - E + F = \chi = 2$. Recently, with the rise of interest in the use of higher-order networks to investigate complex systems, we are witnessing an increase in the use of topology to analyze real-world network science datasets [8, 15, 18, 28]. Topology reveals the structure of higher-order networks at different scales, ranging from the local structure of topological node neighborhoods [24] to the large-scale structure of Topological Data Analysis (TDA) and persistent homology [20, 21]. As we will see in Sections 6 and 7, topology can also play a fundamental role in the study of the dynamics of topological signals, i.e. dynamical variables not only associated with the nodes of the simplicial complex, but also with the links, triangles and general higher-order simplices.

This section aims to introduce the most recent developments in simplicial network topology. In Section 3.2 we will provide a brief introduction to the algebraic topology of simplicial complexes, and discuss how algebraic topology can be used to define the Betti numbers β_m and the Euler characteristic χ of simplicial complexes. In Sections 3.3 and 3.4 we will discuss more in detail the novel topological tools that are available to investigate higher-order network data. Finally, in Section 3.5 we will discuss how topology is related to the community structure of the simplicial complexes.

3.2 A Brief Introduction to Algebraic Topology

3.2.1 Oriented Simplices and the m-Chains

Algebraic topology [19, 66] attributes to each simplex an orientation, and treats the set of all m-dimensional oriented simplices of a simplicial complex as the basis of the linear space of m-chains which are at the foundation of this mathematical subject.

ORIENTED SIMPLICES

A m-dimensional *oriented simplex* α is a set of $m + 1$ nodes

$$\alpha = [v_0, v_1, \ldots, v_m], \tag{3.1}$$

associated with an orientation such that

$$[v_0, v_1, \ldots, v_m] = (-1)^{\sigma(\pi)} [v_{\pi(0)}, v_{\pi(1)}, \ldots, v_{\pi(m)}] \tag{3.2}$$

where $\sigma(\pi)$ indicates the parity of the permutation π.

For instance, the link $\alpha = [r, s]$ has the opposite sign to the link $[s, r]$, i.e.

$$[r, s] = -[s, r]. \tag{3.3}$$

It is good practice to associate each simplex of the simplicial complex and the orientation induced by the node labels, so that

$$\alpha = [v_0, v_1, \ldots, v_m] \tag{3.4}$$

is associated with a positive orientation if

$$v_0 < v_1 < \ldots < v_m. \tag{3.5}$$

In Figure 12 we show a 1-dimensional simplex and a 2-dimensional simplex with orientations induced by the node labels. This practice is convenient when working with higher-order Laplacians because it is possible to prove that the

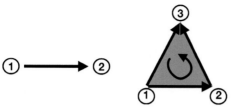

Figure 12 Example of oriented 1-dimensional and 2-dimensional simplices with orientations induced by the node labels.

Source: Reprinted figure with permission from [27] ©Copyright (2020) by the American Physical Society.

spectral properties of these operators are independent of the choice of the node labels as long as the orientation of the simplices is performed in this way.

THE m-CHAINS

Given a simplicial complex, an m-chain $\mathcal{C}_m$ consists of the elements of a free abelian group based on the m-simplices of the simplicial complex. Its elements can be represented as linear combinations of all of the oriented m-simplices

$$\alpha = [v_0, v_1, \ldots, v_m] \tag{3.6}$$

with coefficients in $\mathbb{Z}$.

Therefore, every element $a \in \mathcal{C}_m$ can be uniquely expressed as a linear combination of the basis elements (m-simplices). In Figure 13 we show a simplicial complex and an example of a 1-chain $a \in \mathcal{C}_1$ with

$$a = [1, 2] - [2, 3] + [2, 4]. \tag{3.7}$$

3.2.2 The Boundary Maps

The boundary maps are fundamental operators acting on m-chains.

THE BOUNDARY MAP

The boundary map ∂_m is a linear operator

$$\partial_m : \mathcal{C}_m \to \mathcal{C}_{m-1} \tag{3.8}$$

whose action is determined by the action on each m-simplex of the simplicial complex and is given by

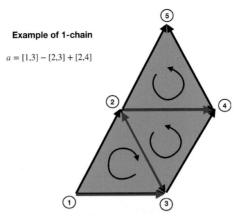

Example of 1-chain

$a = [1,3] - [2,3] + [2,4]$

Figure 13 A simplicial complex whose simplices have an orientation induced by the node labels. On this simplicial complex we highlight in red the 1-chain $a = [1,2] - [2,3] + [2,4]$.

$$\partial_m[v_0, v_1 \ldots, v_m] = \sum_{p=0}^{m}(-1)^p[v_0, v_1, \ldots, v_{p-1}, v_{p+1}, \ldots, v_m]. \quad (3.9)$$

From this definition it follows that the $\mathrm{im}(\partial_m)$ corresponds to the space of $(m-1)$ boundaries and the $\ker(\partial_m)$ is formed by the cyclic m-chains.

Therefore the boundary map maps an m simplex to the $(m-1)$-chain formed by the simplices at its boundary. For instance, we have

$$\partial_1[r, s] = [s] - [r],$$
$$\partial_2[r, s, q] = [r, s] + [s, q] - [r, q]. \quad (3.10)$$

In particular this last expression indicates that the boundary map of a triangle is a cyclic chain formed by the links at its boundary.

THE BOUNDARY OF THE BOUNDARY IS NULL

The boundary operator ∂_m has the topological and algebraic property

$$\partial_m\partial_{m+1} = 0 \ \forall m \geq 1 \quad (3.11)$$

which is usually indicated by saying that *the boundary of the boundary is null.* This implies that

$$\mathrm{im}(\partial_{m+1}) \subseteq \ker(\partial_m). \quad (3.12)$$

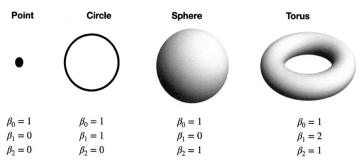

Point	Circle	Sphere	Torus

$\beta_0 = 1$ $\beta_0 = 1$ $\beta_0 = 1$ $\beta_0 = 1$
$\beta_1 = 0$ $\beta_1 = 1$ $\beta_1 = 0$ $\beta_1 = 2$
$\beta_2 = 0$ $\beta_2 = 0$ $\beta_2 = 1$ $\beta_2 = 1$

Figure 14 A point, a circle, a sphere and a torus and their corresponding Betti numbers.

This property follows directly from the definition of the boundary. As an example we have

$$\partial_1 \partial_2 [r, s, q] = \partial_1 ([s, q] - [r, q] + [r, s])$$
$$= [q] - [s] - [q] + [r] + [s] - [r] = 0. \qquad (3.13)$$

3.2.3 Betti Numbers and the Euler Characteristic

The Betti numbers are topological invariants derived from the simplicial complex and correspond, for each $m \geq 0$, to the number of linearly independent m-dimensional cavities in the space. Specifically the Betti number β_0 provides the number of connected components of the simplicial complex, the Betti number β_1 measures the number of 1-dimensional holes, i.e. cycles that are not boundaries of 2-dimensional subsets of the simplicial complex, and so on for higher-order Betti numbers. Betti numbers are not only defined for discrete topological spaces such as simplicial complexes, but they also characterize the topology of continuous topological spaces, such as a point ($\beta_0 = 1, \beta_1 = \beta_2 = 0$), a circle ($\beta_0 = \beta_1 = 1, \beta_2 = 0$), a sphere ($\beta_0 = 1, \beta_1 = 0, \beta_2 = 1$) and a torus ($\beta_0 = 1, \beta_1 = 2, \beta_2 = 1$) (see Figure 14). The Betti numbers are fundamental topological invariants that characterize higher-order networks represented by simplicial complexes. For instance, in Figure 15 we represent the data for a fungi network studied in Ref. [67]. This planar network has a topology that can be studied by performing the clique complex of the data, i.e. interpreting every triangle as a 2-dimensional simplex and studying its Betti number β_1 which characterizes the number of 1-dimensional holes. This topological data analysis provides very important information about this dataset as is already apparent from its planar network representation. Betti numbers

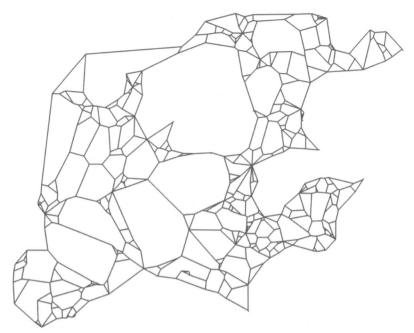

Figure 15 A fungus network. From this dataset it is clearly evident that the clique complex of this dataset will be characterized by a large Betti number β_1 corresponding to the large number of non-triangulated cycles found in this dataset.

Source: Data published in [67].

are becoming a very important tool to understand the topology of higher-order networks, in particular in neuroscience applications [8, 20–23].

In this section we will provide the mathematical definition of Betti numbers using the basic elements of algebraic topology introduced in previous sections. In Section 3.2.2 we have shown that the boundaries of $(m+1)$-chains are cyclic m-chains which belong to $\ker(\partial_m)$, or in other words $\ker(\partial_m) \subseteq \mathrm{im}(\partial_{m+1})$. In intuitive terms these are boundaries of regions of the simplicial complexes that are "filled" by $(m+1)$-dimensional simplices. For instance the boundary of a triangle $[r, s, q]$ is given by the 1-cyclic chain of links at its boundary. However, when the simplicial complex displays m-dimensional cavities, it means that there are m-cyclic chains that do not delimit an $(m+1)$-dimensional region of the simplicial complex that is filled by $(m+1)$-dimensional simplices. This implies that there are m-cyclic chains which do belong to $\ker(\partial_m)$ but do not belong to $\mathrm{im}(\partial_{m+1})$. The set of all cyclic m-chains can be classified according to different homology groups.

THE HOMOLOGY GROUPS

The homology group $\mathcal{H}_m$ is the quotient space

$$\mathcal{H}_m = \frac{\ker(\partial_m)}{\operatorname{im}(\partial_{m+1})}, \tag{3.14}$$

denoting homology classes of m-cyclic chains that are in $\ker(\partial_m)$ and do differ by cyclic chains that are not boundaries of $(m+1)$-chains, i.e. they are in $\operatorname{im}(\partial_{m+1})$.

If follows that $a \in \ker(\partial_m)$ is in the same homology class as $a' = a + b$ where $b \in \operatorname{im}(\partial_{m+1})$. This means that two cyclic m-chains a and a' that only differ by a boundary of an $(m+1)$-chain are in the same homology class. Indeed these two chains will enclose the same number of m-dimensional cavities. However, $a \in \ker(\partial_m)$ and $a' \in \ker(\partial_m)$ will not belong to the same homology class if they enclose a different number of m-dimensional cavities. The total number of m-dimensional cavities is indicated by the Betti number β_m whose algebraic topology definition is given below.

BETTI NUMBERS

The Betti number β_m indicates the number of m-dimensional cavities of a simplicial complex and is given by the rank of the homology group $\mathcal{H}_m$, i.e.

$$\beta_m = \operatorname{rank}(\mathcal{H}_m) = \operatorname{rank}(\ker(\partial_m)) - \operatorname{rank}(\operatorname{im}(\partial_{m+1})). \tag{3.15}$$

The Betti numbers are fundamental topological invariants and as such they are the pillars of topology and TDA, as we will discuss in the next sections.

The Euler–Poincaré formula relates the Betti number to another important topological invariant of simplicial complexes: the Euler characteristic.

THE EULER CHARACTERISTIC AND THE EULER–POINCARÉ FORMULA

The Euler characterisic χ is defined as the alternating sum of the number of m-dimensional simplices, i.e.

$$\chi = \sum_{m \geq 0} (-1)^m s_m, \tag{3.16}$$

where s_m is the number of m-dimensional simplices in the simplicial complex. According to the Euler–Poincaré formula, the Euler characteristic χ of a simplicial complex can be expressed in terms of the Betti numbers as

$$\chi = \sum_{m \geq 0} (-1)^m \beta_m. \tag{3.17}$$

3.2.4 Incidence Matrices

For every dimension m of a simplicial complex we consider an ordered list of its m-simplices and we take this list as a basis for the space of m-chains. With this choice of basis for the m-chains and for the $(m-1)$-chains, the m-boundary operator ∂_m defined in Eq. (3.9) can be represented by the $N_{[m-1]} \times N_{[m]}$ incidence matrix $\mathbf{B}_{[m]}$ of non-zero elements given by

$$\left[\mathbf{B}_{[m]}\right]_{\alpha',\alpha} = (-1)^p, \tag{3.18}$$

for every pair of m-dimensional simplex α and $(m-1)$ dimensional simplex α' such that

$$\alpha = [v_0, v_1 \ldots, v_m], \tag{3.19}$$
$$\alpha' = [v_0, v_1 \ldots, v_{p-1}, v_{p+1} \ldots, v_m]. \tag{3.20}$$

For instance, for the simplicial complex shown in Figure 16 the incidence matrices $\mathbf{B}_{[1]}$ and $\mathbf{B}_{[2]}$ read,

$$
\mathbf{B}_{[1]} =
\begin{array}{c|cccc}
 & [1,2] & [1,3] & [2,3] & [3,4] \\
\hline
[1] & -1 & -1 & 0 & 0 \\
[2] & 1 & 0 & -1 & 0 \\
[3] & 0 & 1 & 1 & -1 \\
[4] & 0 & 0 & 0 & 1
\end{array}
\qquad
\mathbf{B}_{[2]} =
\begin{array}{c|c}
 & [1,2,3] \\
\hline
[1,2] & 1 \\
[1,3] & -1 \\
[2,3] & 1 \\
[3,4] & 0
\end{array}
. \tag{3.21}
$$

In terms of the incidence matrices the relation expressed in Eq. (3.11) can be expressed as

$$\mathbf{B}_{[m]}\mathbf{B}_{[m+1]} = \mathbf{0}, \quad \forall m \geq 1$$
$$\mathbf{B}_{[m+1]}^\mathsf{T}\mathbf{B}_{[m]}^\mathsf{T} = \mathbf{0}, \quad \forall m \geq 1. \tag{3.22}$$

3.2.5 Higher-Order Laplacians and Hodge Decomposition

The graph Laplacian is a fundamental operator that describes diffusion occurring from a node to another node through links. The graph Laplacian matrix

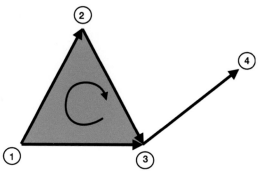

Figure 16 An example of a simplicial complex whose boundary matrices $\mathbf{B}_{[1]}$ and $\mathbf{B}_{[2]}$ are given by Eq. (3.21)

Source: Reprinted with permission from [27].

$\mathbf{L}^{[0]}$ is an $N_{[0]} \times N_{[0]}$ matrix typically defined in terms of the diagonal matrix $\mathbf{K}$ having the degrees of the nodes on the diagonal and the adjacency matrix $\mathbf{A}$ of the network as

$$\mathbf{L}^{[0]} = \mathbf{K} - \mathbf{A}. \tag{3.23}$$

However, the graph Laplacian can be equivalently defined in terms of the incidence matrix $\mathbf{B}_{[1]}$ as

$$\mathbf{L}^{[0]} = \mathbf{B}_{[1]}\mathbf{B}_{[1]}^{\top}. \tag{3.24}$$

This expression can be generalized in order to define higher-order Laplacians $\mathbf{L}_{[m]}$ (also called combinatorial Laplacians) that describe diffusion from an m simplex to another m simplex.

HIGHER-ORDER LAPLACIAN

The higher-order Laplacian operator can be represented as an $N_{[m]} \times N_{[m]}$ matrix. Since for $m > 0$, diffusion from an m simplex to another m simplex can occur either though an $(m-1)$-simplex or though an $(m+1)$-simplex, the higher-order Laplacian $\mathbf{L}_{[m]}$ with $m > 0$ can be decomposed as

$$\mathbf{L}_{[m]} = \mathbf{L}_{[m]}^{down} + \mathbf{L}_{[m]}^{up}, \tag{3.25}$$

with

$$\mathbf{L}_{[m]}^{down} = \mathbf{B}_{[m]}^{\top}\mathbf{B}_{[m]}, \tag{3.26}$$

$$\mathbf{L}_{[m]}^{up} = \mathbf{B}_{[m+1]}\mathbf{B}_{[m+1]}^{\top}. \tag{3.27}$$

The $\mathbf{L}_{[m]}^{down}$ Laplacian describes the diffusion from m-simplices to m-simplices through $(m-1)$-simplices. The $\mathbf{L}_{[m]}^{up}$ Laplacian describes the diffusion from m-simplices to m-simplices through $(m+1)$-simplices. The higher-order Laplacian can be proven to be independent of the orientation of the simplices as long as this orientation is induced by the node labels.

As a consequence of the definition of the higher-order Laplacian we make the following two observations (see Appendix B for details):

- The eigenvector of $\mathbf{L}_{[m]}^{down}$ corresponding to a given eigenvalue $\lambda = \mu^2 > 0$ of $\mathbf{L}_{[m]}^{down}$ is the right eigenvector of $\mathbf{B}_{[m]}$ corresponding to the eigenvalue μ.
- The eigenvector of $\mathbf{L}_{[m]}^{up}$ corresponding to a given eigenvalue $\lambda = \mu^2 > 0$ of $\mathbf{L}_{[m]}^{up}$ is the left eigenvector of $\mathbf{B}_{[m+1]}$ corresponding to the eigenvalue μ.

The main property of the higher-order Laplacians used by topology is that the degeneracy of the zero eigenvalue of the m-Laplacian $\mathbf{L}_{[m]}$ is equal to the Betti number β_m and that their corresponding eigenvectors localize around the corresponding m-dimensional cavity of the simplicial complex. Therefore, the higher-order Laplacians with $m > 0$ are not guaranteed to have a zero eigenvalue as, for some $m > 0$, simplicial complexes with $\beta_m = 0$ exist.

By investigating the properties of higher-order Laplacians it is possible to decompose the space Ω^m of all m-chains according to Hodge decomposition (see Appendix B for details), which can be summarized as

$$\Omega^m = \text{img}(\mathbf{B}_{[m]}^{\top}) \oplus \ker(\mathbf{L}_{[m]}) \oplus \text{img}(\mathbf{B}_{[m+1]}). \tag{3.28}$$

The Hodge decomposition implies that $\mathbf{L}_{[m]}$, $\mathbf{L}_{[m]}^{up}$ and $\mathbf{L}_{[m]}^{down}$ are commuting and can be diagonalized simultaneously. In this basis these matrices have the block structure

$$\mathbf{U}^{-1}\mathbf{L}_{[m]}\mathbf{U} = \begin{pmatrix} \mathbf{D}_{[m]}^{down} & \mathbf{0} & \mathbf{0} \\ \mathbf{0} & \mathbf{0} & \mathbf{0} \\ \mathbf{0} & \mathbf{0} & \mathbf{D}_{[m]}^{up} \end{pmatrix}, \tag{3.29}$$

$$\mathbf{U}^{-1}\mathbf{L}_{[m]}^{down}\mathbf{U} = \begin{pmatrix} \mathbf{D}_{[m]}^{down} & \mathbf{0} & \mathbf{0} \\ \mathbf{0} & \mathbf{0} & \mathbf{0} \\ \mathbf{0} & \mathbf{0} & \mathbf{0} \end{pmatrix}, \tag{3.30}$$

$$\mathbf{U}^{-1}\mathbf{L}_{[m]}^{up}\mathbf{U} = \begin{pmatrix} \mathbf{0} & \mathbf{0} & \mathbf{0} \\ \mathbf{0} & \mathbf{0} & \mathbf{0} \\ \mathbf{0} & \mathbf{0} & \mathbf{D}_{[m]}^{up} \end{pmatrix}, \tag{3.31}$$

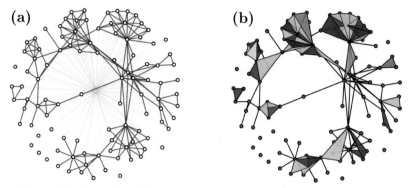

Figure 17 The node neighborhood of a node r formed by considering the subgraph of the original network formed by all the neighbors of node r and their connections (panel a), and performing the clique complex of this subgraph (panel b).

Source: Reprinted with permission [68] from [24].

where $\mathbf{D}_{[m]}^{down}$ and $\mathbf{D}_{[m]}^{up}$ indicate diagonal matrices. Therefore an eigenvector in the ker of $\mathbf{L}_{[m]}$ is also in the ker of both $\mathbf{L}_{[m]}^{up}$ and $\mathbf{L}_{[m]}^{down}$. An eigenvector corresponding to an non-zero eigenvalue of $\mathbf{L}_{[m]}$ is either a non-zero eigenvector of $\mathbf{L}_{[m]}^{up}$ or a non-zero eigenvector of $\mathbf{L}_{[m]}^{down}$.

3.3 Topological Data Analysis of Simplicial Complexes

3.3.1 Topological Clustering of Node Neighborhoods

Topology can be used to shed new light on the local properties of network datasets. It is well known that the local clustering of a network can be quantified by the clustering coefficient [69]. However, the study of node neighborhoods (also called ego-centered networks) can be enriched if one investigates these subgraphs with topological methods [24]. In particular one can consider the node neighborhood of a node r formed by the clique complex of a subgraph of the original network formed by all the neighbors of node r and their connections (see Figure 17). Therefore, the node neighborhood of r excludes node r together with all the links incident to it. The number of nodes in a node neighborhood constructed in this way is equal to the number of links of node r, i.e. its degree. The density of links of the node neighborhood can be easily shown to be equal to the local clustering coefficient of the node r. A topological analysis of node neighborhoods consists in the computation of their Betti numbers. Interestingly a large-scale statistical study of node neighborhoods [24] revealed that nodes with the same degree and the same local clustering coefficient can have node neighborhoods with very different topologies, despite the fact that these node neighborhoods have the same number of nodes and density of links. Therefore the topological analysis of node neighborhoods reveals important differences in the local clustering of networks that go beyond the information captured by the clustering coefficient.

3.3.2 Persistent Homology

Applied topology can also be used very successfully to characterize the large-scale properties of higher-order networks by means of persistent homology. Persistent homology is a pivotal method in Topological Data Analysis that has been used on a variety of datasets, including point clouds and networks, representing a large variety of data coming from very different applications ranging from neuroscience to cancer biology [8, 16, 18, 19]. Persistent homology quantifies relevant topological features of data. In particular, it investigates the properties of the data as a function of a parameter that allows the dataset to be represented as a series of simplicial complexes that grow by the subsequent addition of simplices, called a filtration. On point clouds the filtration is typically performed by connecting all the nodes at a distance of at least δ, and performing the clique complex of the resulting network for different values of δ. For data that already has a network structure, a powerful way to define a filtration [20, 21, 70] is to induce the filtration using the weights of the links. Specifically one can consider the filtration defined by considering only the links whose weight is below δ and performing the clique complex of the resulting network for different values of δ going from zero to the maximum weight of the links of the network. Once a filtration is defined, persistent homology proceeds by computing the homology of each simplicial complex in the filtration, i.e. corresponding to a given value of the parameter δ. The result of persistent homology can be visualized by a barcode that indicates the range of values of δ in which a given m-dimensional homology class is observed (see Figure 18). For $m = 0$, each line corresponds to a different connected component. So typically for small values of δ one observes many components and as the value of δ increases the number of components decreases. For $m > 0$ each bar of the barcode corresponds to an m-dimensional cavity. Persistent homology is based on the idea that the most salient features of the original dataset are the persistent features that are observed for many different values of the tuning parameter δ and correspond to longer bars in the persistent homology barcode. Persistent homology is currently one of the pillars of TDA and has been shown to be a fundamental tool for analyzing higher-order networks, as it can reveal large-scale topological structure of data that is not detected by more traditional

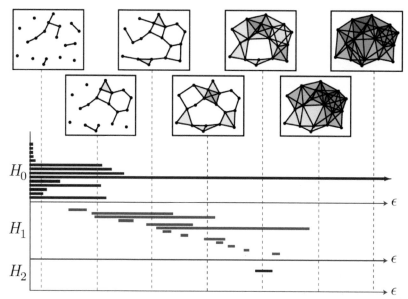

Figure 18 An example of a topological filtration and its corresponding persistent homology barcode. Here H_0 refer to the connected component and H_1 refer to the cycles and H_2 refer to the 2-dimensional cavities of the simplicial complex.

Source: Reproduced with permission from Ref. [16] ©2007 Robert W. Ghrist.

network science measures based on statistical and combinatorial properties of the network.

3.4 Clique Communities and Motif Conductance

3.4.1 Clique Communities and m-Connectedness

Many networks are characterized by displaying a significant number of motifs, including cliques, polytopes and even more general subgraphs such a bipartite cliques that constitute the building blocks of their meso-scale structure. In fact, in real networks motifs often overlap and can be very efficiently used to infer their community structure. This phenomenon is exploited by both the clique community detection algorithm and the community detection algorithm using motif conductance. In clique community detection [26] the goal is to decompose a network onto clusters based on the higher-order connectivity of their clique complex. The clique communities of a network correspond to the mathematical notion of *m*-connected components of the clique complex of the network.

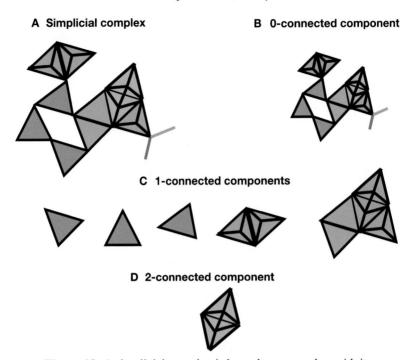

Figure 19 A simplicial complex is here shown together with its decomposition into m-connected components.

---

m-CONNECTED COMPONENTS

Two simplices α and α' are m-connected if there is a sequence of simplices $(\alpha, \alpha_1, \alpha_2, \ldots, \alpha_s, \alpha')$ such that any two consecutive simplices have at least a common m-dimensional face. A simplicial complex is m-connected if any two simplices of dimension greater than or equal to m are m-connected.

In Figure 19 we show the decomposition of a simplicial complex into its m-connected components. The $(m + 2)$-clique communities of a network [26] are the 1-skeleton of the m-connected components of the clique complex of the network. Any simplex of the considered clique complex is adjacent (m-connected) to another simplex if the two simplices have a common m-dimensional face. The resulting m-connected components form the community structure detected by the algorithm. Most notably, the detected communities might overlap on m'-dimensional simplices with $m' < m$. This is a significant property of this community detection algorithm, as in different real contexts it is important to allow communities to overlap. For instance an interesting dataset in which it is important to consider overlapping communities is the word-association network [26]. In the word-association network words are connected by links if they have been associated with each other in a questionnaire presented to a set of people. The clique communities represent words with close meanings associated with higher-level concepts. In this context the English word "bright" can be shown to be associated with the communities related to the concept of color, intelligence and astronomical terms. The clique communities can be detected efficiently in networks by the C-Finder algorithm.

3.4.2 Motif Conductance

More recently, a community detection algorithm based on motif conductance has been proposed [25]. This algorithm is based on a network decomposition on motifs that can be cliques but also more general motifs $\mathcal{H}$. The algorithm generates a non-overlapping community decomposition of the network where each node belongs to a single cluster.

The algorithm is a greedy algorithm based on spectral clustering that aims to minimize the *motif conductance* $\phi_{\mathcal{H}}$ by finding the best bipartition of the network formed by a subgraph H and its complement $\bar{H}$, iteratively.

The motif conductance $\phi_{\mathcal{H}}(H)$ is a function associated with the bisection of the network into a subgraph H and its complement $\bar{H}$, defined as

$$\phi_{\mathcal{H}}(H) = \frac{\text{cut}_{\mathcal{H}}(H, \bar{H})}{\min[|H|_{\mathcal{H}}, |\bar{H}|_{\mathcal{H}}]}, \tag{3.32}$$

where $\text{cut}_{\mathcal{H}}(H)$ indicates the number of motifs $\mathcal{H}$ that are cut by the partition, i.e. that have a fraction of nodes belonging to H with the rest belonging to $\bar{H}$. Moreover, in Eq. (3.32) $|H|_{\mathcal{H}}$ and $|\bar{H}|_{\mathcal{H}}$ are given by

$$|H|_{\mathcal{H}} = \sum_{r \in H} \kappa_r,$$

$$|\bar{H}|_{\mathcal{H}} = \sum_{r \in \bar{H}} \kappa_r \tag{3.33}$$

where κ_r indicates the number of motifs $\mathcal{H}$ incident to node r. While the optimization of the motif conductance is a hard combinatorial problem, the greedy algorithm proposed in Ref. [25] can be efficiently applied to networks of sizes up to millions of nodes.

3.5 Emergent Community Structure

Growing simplicial complex models can start from very simple rules, such as the rule of triadic closure that is inspired by the observation that in social networks the friend of a friend is likely to be a friend, i.e. triplets of connected nodes are likely to close a triangle. Models of triadic closure have been investigated by several authors since the early days of network science. Here we discuss one of the most simple versions and we interpret it as a model of growing simplicial complexes:

SIMPLICIAL COMPLEX MODEL ENFORCING TRIADIC CLOSURE [71]

At time $t = 1$ the 2-dimensional simplicial complex has a skeleton which is a finite connected network with $n_0 > 2$ nodes. The model evolves by the addition of new links and triangles according to the following rules:

(1) GROWTH: At every timestep $t > 1$ we add a new node with z links (connected to the nodes already present in the system).

(2) TRIADIC CLOSURE: The first link is attached to a random node, each of the other links is attached to a random neighbor of the initial node and closes a triangle with probability p; with probability $1 - p$ it is connected instead to a node of the network chosen randomly among all the nodes of the simplicial complex.

The model is summarized in Figure 20 (panels a, b).

The local rule determining the simplicial complex evolution leads, as long as z is finite, to a simplicial complex whose 1-skeleton is a network with a broad degree distribution (but not power-law as widely believed) – see [71] for details. Most importantly Ref. [71] has shown that the local rules enforcing triadic closure are responsible for a very rich meso-scale structure of the network skeleton. Indeed, the mechanism of triadic closure induces the emergence of communities that are dynamically generated and evolve in time (see Figure 20c). Note that the model is defined in such a way that the simplicial complex generated by the model is always 2-dimensional (for $p > 0$), i.e. the largest clique of the network is a triangle; however, the meso-scale community that emerges from the evolving simplicial complex model involves a number of nodes that can be much larger than three.

4 Simplicial Network Geometry

4.1 Discrete Manifolds

Simplicial complexes allow the definition of a simplicial network geometry providing a very powerful tool to analyze discrete higher-order network data and to investigate dynamics defined on them. In the geometry of continuous metric spaces, manifolds, including $d = 2$ dimensional surfaces and their

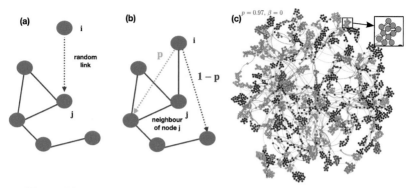

Figure 20 Panels (a) and (b) represent schematically the growth of the network skeleton of the model of a growing simplicial complex with triadic closure. Panel (c) shows the emergent community structure obtained considering the parameters $z = 2$ and $p = 0.95$.

Source: Reprinted figure with permission from [71] ©Copyright (2014) by the American Physical Society.

higher-dimensional equivalents, play a very special role. However, in network science, simplicial complexes that are discrete manifolds as well as simplicial complexes that are not discrete manifolds play a very relevant role in the underlying structure of complex systems. Therefore, when geometrical concepts are used to characterize higher-order networks it is important to make a distinction between simplicial complexes that are discrete manifolds and simplicial complexes that are not. In this section we will provide the combinatorial conditions that a general simplicial complex needs to satisfy in order to be a discrete manifold. Before giving these conditions it is useful to introduce the *incidence number* associated with the $(d-1)$-faces of pure d-dimensional simplicial complexes [29].

INCIDENCE NUMBER

The incidence number n_α of a $(d-1)$-dimensional face α of a pure d-dimensional simplicial complex $\mathcal{K}$, is given by its generalized degree minus one, i.e.

$$n_\alpha = k_{d,d-1}(\alpha) - 1. \tag{4.1}$$

Therefore for each $(d-1)$-dimensional face $\alpha \in S_{d-1}(\mathcal{K})$, we have $n_\alpha \geq 0$.

We are now in the position to provide the combinatorial conditions that a simplicial complex needs to satisfy in order to be a discrete manifold.

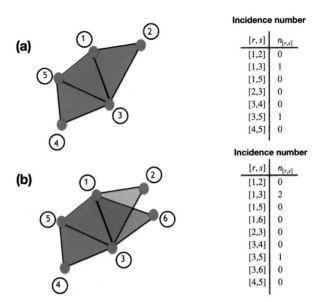

Figure 21 A 2-dimensional discrete manifold is shown together with the list of the incidence numbers of its links (panel a). A 2-dimensional simplicial complex that is not a manifold is shown together with the list of the incidence number of its links (panel b). The simplicial complex in panel (b) deviates from a manifold because it has a "fin" on the link $[1, 3]$. This branched structure is revealed by the incidence number $n_{[1,3]} = 2$.

COMBINATORIAL CONDITIONS FOR DISCRETE MANIFOLDS

A discrete manifold $\mathcal{M}$ of dimension d is a pure simplicial complex that satisfies the following two conditions:

- it is $(d-1)$-connected;
- every two d-simplices α, α' belonging to the simplicial complex $\mathcal{K}$, (i.e. $\alpha, \alpha' \in S_{d-1}(\mathcal{M})$ either overlap on a $(d-1)$-face of $\mathcal{K}$, i.e. $\alpha \cap \alpha' \in S_{d-1}(\mathcal{K})$ or do not overlap, i.e. $\alpha \cap \alpha' = \emptyset$.
- all $(d-1)$-faces $\alpha \in S_{d-1}(\mathcal{M})$ have an incidence number $n_\alpha \in \{0, 1\}$.

The last condition, imposing that all the incidence numbers of the $(d-1)$-faces of the simplicial complex have value $n_\alpha \in \{0, 1\}$ guarantees that there are no "fins" in the simplicial complexes (see Figure 21 for a visual representation of the implications of having $n_\alpha > 1$). In a manifold $\mathcal{M}$ we will distinguish between the bulk $\mathcal{B}$ and the boundary $\mathcal{A}$ including all the simplices that are not in the bulk.

> **BOUNDARY AREA AND BULK OF A DISCRETE MANIFOLD**
> The boundary $\mathcal{A}$ of a d-dimensional discrete manifold $\mathcal{M}$ is formed by the set of all $(d-1)$-dimensional faces $\alpha \in \mathcal{M}$ with incidence number $n_\alpha = 0$ and by all their faces. The area A is the number of $(d-1)$-dimensional faces in the boundary $\mathcal{A}$. The bulk $\mathcal{B}$ of a discrete manifold $\mathcal{M}$ is formed by the set of all the faces that are not in the boundary $\mathcal{A}$.

4.2 Curvature

4.2.1 Regge Curvature

In differential geometry a crucial role is played by the curvature of continuous manifolds. When treating simplicial network geometry an important problem is to formulate a good definition for the curvature of discrete simplicial complexes, which can be treated as the discrete counterpart of the notion used in differential geometry. Different definitions have been proposed in the literature [72–75] of which the Regge curvature is one of the most popular. The Regge curvature [76, 77] is an important definition of curvature for discrete manifolds. This definition has been given in [76] in order to propose a discrete version of general relativity with the correct continuum limit.

> **REGGE CURVATURE**
> The Regge curvature [75] is associated wtih each $(d-2)$-dimensional face $\alpha \in S_{d-2}(\mathcal{M})$ of a discrete d-dimensional manifold $\mathcal{M}$. The Regge curvature R_α for a face $\alpha \in S_{d-2}(\mathcal{M})$ is defined as
> $$R_\alpha = \begin{cases} 2\pi - \theta_\alpha & \text{if } \alpha \in \mathcal{B}, \\ \pi - \theta_\alpha & \text{if } \alpha \in \mathcal{A}, \end{cases} \tag{4.2}$$
> where θ_α is the sum of all dihedral angles of the d-dimensional simplices incident to the face α.

It follows that if the discrete manifold is formed by a set of geometrically identical d-simplices, the Regge curvature is simply related to the generalized degree of the $(d-2)$-faces, i.e.

$$R_\alpha = \begin{cases} 2\pi - \theta_0 k_{d,d-2}(\alpha) & \text{if } \alpha \in \mathcal{B}, \\ \pi - \theta_0 k_{d,d-2}(\alpha) & \text{if } \alpha \in \mathcal{A}, \end{cases} \tag{4.3}$$

where θ_0 indicates the dihedral angle of each d-simplex. An instructive example of how the Regge curvature works is provided by the $d = 2$ case in which all the 2-dimensional simplices of the manifold are equilateral triangles of dihedral angle $\theta_0 = \pi/3$ (60° degrees). In this case the Regge curvature is associated wtih the nodes of the manifold. If a node is in the bulk $\mathcal{B}$ of the manifold, the Regge curvature has a sign that depends on the number of equilateral triangles incident to it. If the node is incident to six equilateral triangles, the curvature of the node is null, as locally the manifold can be embedded in $\mathbb{R}^2$, if, however, the node is incident to more than six equilateral triangles, the curvature of the node is negative; inversely, if the node is incident to less than six equilateral triangles the curvature of the node is positive. The Regge curvature was proposed by Regge in order to formulate a discrete version of general relativity. In this framework the Einstein–Hilbert action is substituted by the Regge action.

> ### REGGE ACTION
> The Regge action S is given by
> $$S = \sum_{\alpha \in S_{d,d-2}(\mathcal{M}} V_\alpha R_\alpha \qquad (4.4)$$
> where V_α is the measure associated with the $d-2$ face α, for $d = 2$ $V_\alpha = 1$ for each node of the manifold, for $d = 3$, V_α indicates the length of the link α, for $d = 4$, V_α indicates the area of the triangle α.

The Regge action is known to reduce to the Einstein–Hilbert action in the continuum limit. For this reason the Regge action is adopted for discrete numerical investigations of classical general relativity, and it is also a reference for quantum gravity approaches including Causal Dynamical Triangulation [78] and Regge Calculus, among others.

The Regge curvature satisfies the Gauss–Bonnet theorem that was originally derived for continuous manifolds.

> ### GAUSS–BONNET THEOREM
> According to the Gauss–Bonnet theorem the sum over the Regge curvatures R_α of all the $(d-2)$-faces α of the discrete d-dimensional manifold $\mathcal{M}$ is a topological invariant proportional to the Euler characteristic χ of the manifold, i.e.
> $$\sum_{\alpha \in S_{d,d-1}(\mathcal{M})} R_\alpha = 2\pi\chi. \qquad (4.5)$$

Therefore, geometrical deformations of the simplicial complex such as stretching or twisting do modify the dihedral angles of the simplicial complex and do affect the Regge curvature locally. However, globally the sum of the Regge curvatures is not affected by geometrical deformations and is proportional to the topological invariant χ.

4.2.2 Gromov δ-Hyperbolicity

Although the Regge curvature is widely adopted in simplicial network geometry, finding alternative definitions of curvature for networks and simplicial complexes is a very active research field at the moment. Alternative definitions include the popular Ollivier–Ricci curvature [72–74] and the Forman curvature [75] that we will not cover here due to space limitations. For physicists interested in the continuum limit, a limitation of the Regge curvature is that it is a scalar and does not constitute a counterpart for the Ricci curvature tensor. For network scientists an important limitation of the Regge curvature is that it applies only to discrete manifolds and does not provide insights for measuring the curvature for simplicial complexes that are not manifolds. In order to address the latter problem, one pivotal contribution is the definition of the Gromov δ-hyperbolicity [79–82]. This definition aims to provide a global quantification of the overall hyperbolicity of a network, which does not necessarily need to be the skeleton of a discrete manifold. Therefore the definition of the Gromov δ-hyperbolicity can be very useful to mathematically characterize the geometry of many real biological and technological networks. As hyperbolic spaces are known to have slim triangles, the Gromov δ-hyperbolicity definition quantifies how *slim* the triangles formed by the shortest paths among three nodes of the network are. A path $\mathcal{P}_{rs}$ between node r and s is an alternate sequence of nodes and adjacent links starting at node r and ending at node s. The length of a path is given by the number of links traversed by the path. Among all the paths between two nodes r and s, the shortest paths are the paths of shortest length. Their length is also called the distance between the nodes r and s. Let us indicate as the δ-neighborhood of a path, the set of all the nodes at a distance less than or equal to δ from any node belonging to the path, and all the links among them.

GROMOV δ-HYPERBOLICITY

A network is said to be δ-hyperbolic if it obeys the δ-slim property, i.e. if there is a $\delta > 0$ such that for any triple of nodes r, s, q connected by the shortest paths $\mathcal{P}_{rs}, \mathcal{P}_{sq}, \mathcal{P}_{rq}$ the union of the δ-neighborhood of any pair of shortest paths, say $N_\delta(\mathcal{P}_{rs}) \cup N_\delta(\mathcal{P}_{sq})$ includes nodes belonging to the third path, i.e. $\mathcal{P}_{rq}$.

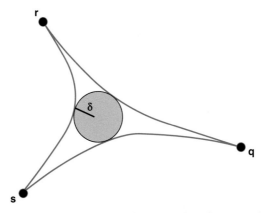

Figure 22 Schematic representation of three nodes of a network skeleton of a simplicial complex and the "triangle" formed by the shortest paths connecting them. Networks that are δ-hyperbolic have "slim-triangles" that obey Gromov δ-hyperbolicity with a small value of δ.

In Figure 22 we provide a schematic description of a slim triangle. Note that according to the definition of δ-hyperbolicity, all trees are δ-hyperbolic when $\delta = 0$. The investigation of the δ-hyperbolicity of a network is a rather active field of research from which it emerges that many real networks are actually δ-hyperbolic with a δ that is small compared with the network diameter [81, 82].

4.3 The Hausdorff Dimension of Network Models

A network is a metric space in which the distances between a pair of nodes is given by the shortest length of any path connecting the two nodes. If we have access to a measure of the "length" of a link, these shortest distances can be real numbers; however, in the majority of cases the hopping distance between two nodes is considered, which measures the number of links in the shortest path between them. The Hausdorff dimension is a very fundamental tool to investigate the properties of continuous metric spaces. This concept can be adapted to characterize network models in which we can tune the total number of nodes N.

HAUSDORFF DIMENSION OF NETWORK MODELS
Given a class of network models, such as d-dimensional lattices or ensembles of networks with a given degree distribution, for which we can

consider a series of models with increasing network size N, the Haudorff dimension d_H scales for $N \gg 1$

$$D \sim N^{1/d_H},$$
(4.6)

where D is the diameter of the network model with N nodes, i.e. the maximum among all the shortest distances between any two pair of nodes in the network.

It follows that for a d-dimensional lattice the Hausdorff dimension is

$$d_H = d,$$
(4.7)

while for small-world networks [68] where the diameter D scales with the logarithm of the total number of nodes of the network (i.e. $D = \mathcal{O}(\ln N)$) we put

$$d_H = \infty.$$
(4.8)

4.4 The Spectral Dimension

Despite most real-world networks and real-world higher-order networks being small world, and having an infinite Hausdorff dimension, some higher-order networks with relevant geometrical properties can display a finite spectral dimension d_S. The spectral dimension d_S of the network is the dimension perceived by a diffusion process taking place on the network [34]. As such, the spectral dimension is considered an important probe for characterizing different approaches to quantum gravity [78, 83, 84].

On finite, Euclidean d-dimensional lattices the spectral dimension d_S is equal to the Hausdorff dimension d. However, on more complex geometries, such as fractals, the spectral dimension typically differs from the spectral Hausdorff dimension. The spectral dimension is a concept originating from the study of elastic media (including the investigation of phonons in condensed-matter materials) concerned with the density of modes of vibrations of different lattices. The spectral dimension of a network can be revealed by the spectral properties of the graph Laplacian $\mathbf{L}_{[0]}$. The graph Laplacian defined in Eqs. (3.23) and (3.24) is a semi-positive definite $N \times N$ matrix whose eigenvalues λ_r can be sorted in increasing order,

$$0 = \lambda_1 \leq \lambda_2 \leq \lambda_3 \leq \dots \lambda_N.$$
(4.9)

The Laplacian is key for describing diffusion and the Kuramoto model on networks and constitutes a natural link between topology and dynamics. An important eigenvalue of the graph Laplacian is the smallest non-zero eigenvalue, also called the *Fiedler eigenvalue* or alternatively the *spectral gap*. On a finite and connected network the Fiedler eigenvalue coincides with λ_2 and the inverse of λ_2 defines the typical time-scale for relaxation of the diffusion process to its steady state. The larger λ_2, the faster the relaxation to equilibrium. Expanders are graphs that have a large spectral gap (such as random graphs). In geometrical network models (such as lattices) in which

$$\lambda_2 \to 0 \quad \text{as} \quad N \to \infty, \tag{4.10}$$

we say that the *spectral gap closes*.

SPECTRAL DIMENSION

Geometrical networks (such as a skeleton of geometrical simplicial complexes), in which the spectral gap closes, display a finite *spectral dimension* d_S if for $\lambda \ll 1$, the density of eigenvalues $\rho(\lambda)$ scales as

$$\rho(\lambda) \simeq C\lambda^{d_S/2-1}, \tag{4.11}$$

where C is a constant.

We observe here that, in the presence of a finite spectral dimension, the cumulative distribution $\rho_c(\lambda)$ evaluating the density of eigenvalues $\lambda' \leq \lambda$, for $\lambda \ll 1$, follows the scaling

$$\rho_c(\lambda) \simeq C'\lambda^{d_S/2}, \tag{4.12}$$

where C' is a constant. In the presence of a finite spectral dimension it is possible to evaluate the scaling with the network size of the smallest non-zero eigenvalue λ_2 of a connected network (also called the the Fiedler eigenvalue) by imposing that

$$\rho_c(\lambda_2) = \frac{1}{N}, \tag{4.13}$$

i.e. the eigenvalue λ_2 is the smallest non-zero eigenvalue. From this relation and the scaling of the cumulative density of eigenvalues we get

$$\lambda_2 \propto N^{-2/d_S}, \tag{4.14}$$

predicting the scaling of the Fidler eigenvalue λ_2 with N as $N \to \infty$ and $\lambda_2 \to 0$. The spectral dimension d_S of the network determines the diffusion properties of the network. For a d-dimensional Euclidean lattice the spectral dimension d_S is equal to the Hausdorff dimension d_H, i.e. $d_S = d_H = d$ (see Appendix C). In general, the spectral dimension d_S of a network is related to the Hausdorff dimension d_H by the inequalities [85, 86]

$$d_H \geq d_S \geq 2\frac{d_H}{d_H + 1}. \tag{4.15}$$

Therefore, for small-world networks, which have an infinite Hausdorff dimension $d_H = \infty$, it is only possible to have a finite spectral dimension $d_S \geq 2$. If some regularity conditions are met, the spectral dimension d_S of the graph Laplacian $\mathbf{L}_{[0]}$ is the same as the spectral dimension of the normalized graph Laplacian $\tilde{\mathbf{L}}$ of elements $\tilde{L}_{rs} = \delta(r, s) - A_{rs}/k_r$ describing the stochastic matrix determining the possible transition of a random walker moving on the network skeleton [87]. The spectral dimension d_S of the normalized graph Laplacian characterizes the return-time distribution of the random walker. Indeed, a random walker starting at time $t = 0$ from a random node has probability $\pi_0(t)$ of returning to the initial node after a time t with $\pi_0(t)$ given by

$$\pi_0(t) = \int d\lambda \rho(\lambda) e^{-\lambda t} \propto t^{-d_S/2}. \tag{4.16}$$

In other words, network skeletons with a finite spectral dimension have a return-time probability that decays with time as a power-law. The exponent of this power-law decreases with the spectral dimension d_S. Therefore smaller spectral dimensions correspond to slower relation dynamics of the diffusion process of the random walker.

4.5 Higher-Order Spectral Dimension

Very recently it has been shown in [33, 88] that d-dimensional simplicial complexes with a distinct geometrical nature might display not just a single spectral dimension d_S, indicating the spectral dimension of the graph Laplacian $\mathbf{L}_{[0]}$, rather they can be chacterized by a vector of spectral dimensions

$$\mathbf{d}_S = (d_S^{[0]}, d_S^{[1]}, \dots, d_S^{[d-1]}), \tag{4.17}$$

where the spectral dimension $d_S^{[0]}$ indicates the spectral dimension of the graph Laplacian $\mathbf{L}_{[0]}$, and the spectral dimension $d_S^{[m]}$ with $m > 0$ indicates the spectral dimension of the m-order up-Laplacian $\mathbf{L}_{[m]}^{up}$. Higher-order Laplacians describe different types of diffusion processes that can be defined on a simplicial complex: going from node to node passing through links or from link to link passing through triangles and so on. The higher-order spectral dimension $d_S^{[m]}$ characterizes the non-trivial part of the spectrum of the higher-order (m-order) up-Laplacians. Indeed the m-order up-Laplacians $\mathbf{L}_{[m]}^{up}$ with $m > 0$ have many zero eigenvalues. Therefore the higher-order spectral dimension can be extracted by the power-law scaling of the density of non-zero eigenvalues $\rho_{[m]}^{up}(\lambda)$, i.e.

$$\rho_{[m]}^{up}(\lambda) \propto \lambda^{d_S^{[m]}/2-1}, \qquad\qquad (4.18)$$

for $\lambda \ll 1$. The presence of a vector of spectral dimensions has been revealed only recently and demonstrates the fact that the same simplicial complex higher-order diffusion processes of different orders m can co-exist and can have distinct properties, such as a very different return-time probability distribution [33].

The higher-order spectral dimensions have been observed in Network Geometry with Flavor (models of emergent geometry we will discuss in Section 5) [33] and on their deterministic counterparts [88]. On deterministic simplicial complexes such as Apollonian and pseudofractal simplicial complexes (whose definition is given in Appendix G) it is possible to predict the vector of higher-order spectral dimensions using the renormalization group [88]. As the spectral dimension has been used to probe the universal properties of different quantum gravity approaches [83, 84] it would be interesting to see if a further exploration of their higher-order spectral dimension can distinguish between them.

5 Emergent Geometry

5.1 The Quest for Emergent Geometry

Many complex networks (simple and high-order) have a rich underlying network geometry, ranging from fibers in the brain [89] and biological transportation networks [90, 91] to transcription and metabolic networks not explicitly embedded into a metric space [81, 92].

Several fundamental questions arise from the study of these simple and higher-order networks: What are the mechanisms responsible for the emergence of a distinct geometrical nature of a networked structure? Is it possible to generate discrete network geometries starting from models that are purely combinatorial, or rather should discrete geometry be exclusively considered as the result of dynamics on a latent space? Are there natural priors for the distribution of nodes in the latent space?

In order to answer these questions, in this section we address the problem of *emergent geometry*, while an alternative way of tackling these questions is by using information theory [93]. Emergent geometry refers to the quest for models able to generate networks or simplicial complexes with remarkable geometrical properties starting from non-equilibrium or equilibrium models that do not use any information about this underlying discrete geometry. The quest for models of emergent geometry (also called pre-geometry) has its origins in quantum gravity, where a very fundamental problem is determining how the geometry of continuous space-time emerges from the discrete (or discretized) structure that space-time has at the Planck scale.

The quest for emergent geometry is a very interesting subject at the frontier between mathematics and physics, as it is related at the same time to network geometry, quantum gravity and information theory. As Penrose wrote [94]: *My own view is that ultimately physical laws should find their most natural expression in terms of essentially combinatorial principles,[...]. Thus, in accordance with such a view, should emerge some form of discrete or combinatorial spacetime.* These ideas have been very important for the development of different quantum gravity approaches that assume a discrete nature of quantum space-time.

Emergent geometry is also of significant relevance for network theory and its interdisciplinary applications because it can be used to explore the nature of network geometries that emerge spontaneously from simple combinatorial rules. For instance, it can provide a theoretical framework to characterize how complexity might emerge spontaneously as the outcome of a non-equilibrium evolution of simplicial network geometries. Moreover, it can provide a very rich ground for proposing suitable priors for algorithms that aim at embedding networks and simplicial complexes into metric spaces.

Interestingly, network theory progress on the problem of emergent geometry, developed to understand complexity and biological systems, could bring new insight on some aspects of quantum gravity. In the words of Lee Smolin [95]: *A theory of quantum cosmology cannot be logically consistent if it does not describe a complex universe.*

In this section we discuss models of emergent geometry of simplicial complexes of any topological dimension d. We will embrace a non-equilibrium approach for network evolution by the subsequent addition of simplices to the simplicial complex. We will start by discussing a simple 2-dimensional model of emergent geometry [31] able to generate discrete manifolds and scale-free simplicial complexes as well. We will then move and characterize a very general framework for emergent geometry called Network Geometry with Flavor (NGF). This model generates simplicial complexes of any dimension d which have very distinct statistical and combinatorial properties and display a hyperbolic geometry that can be explored using the δ-hyperbolicity criterion and its relation to hyperbolic tessellations of $\mathbb{H}^d$. Interestingly, NGF also displays emergent community structure and emergent spectral dimensions of the graph Laplacian and of higher-order Laplacians. The last part of this section discusses NGF with fitness of faces, characterizing the intrinsic ability of each face of the simplicial complex to attract new d-dimensional simplices. This variant of the NGF model presents a surprising phenomenon, as the statistical properties of faces of dimension $0 \leq m < d$ are captured by Fermi–Dirac, Boltzman and Bose–Einstein distributions, depending on the dimensions d, m and on the additional parameter s (the flavor). NGF with fitness of the faces reduces in the limit $d = 1, s = 1$ to the Bianconi–Barabási model [96] which displays the Bose–Einstein condensation in complex networks. This section concludes with a discussion of the topological phase transitions displayed by NGF with fitness of the faces for general values of the flavor s, extending the Bose–Einstein phase transition occurring for the Bianconi–Barabási model.

The statistical properties of NGF discussed here have also been confirmed by a pure mathematics paper [97]. The code for every model discussed in this section can be found in the repository [50].

5.2 Emergent 2-Dimensional Simplicial Network Geometry

Let us start by revealing the fundamental mechanism for emergent geometry in dimension $d = 2$ by discussing the non-equilibrium simplicial complex model proposed in Ref. [31]. The model describes the non-equilibrium evolution of a simplicial complex constructed by gluing triangles along the links of a simplicial complex subsequently in time. During their evolution the simplicial complexes need to satisfy a simple combinatorial rule: *at every point in time every link of the simplicial complex must be incident to at most $\bar{k}$ triangles with $\bar{k} > 1$.* We classify links $[r, s]$ as *unsaturated* and *saturated* depending on the value of the auxiliary variable ρ_{rs} defined as

$$\rho_{rs} = \begin{cases} 0 & \text{if} \quad k_{2,1}([r, s]) < \bar{k}, \\ 1 & \text{if} \quad k_{2,1}([r, s]) = \bar{k}. \end{cases} \tag{5.1}$$

Therefore for each link $[r, s]$ there are two possibilities:

- if $\rho_{rs} = 0$ the link is *unsaturated*, i.e. fewer than $\bar{k}$ triangles are incident on it;
- if $\rho_{rs} = 1$ the link is *saturated*, i.e. the number of incident triangles is given by $\bar{k}$.

We define two processes characterizing the different topological moves that can result from the addition of a single triangle to the simplicial complex. These moves, schematically represented in Figure 23, define the model for 2-dimensional emergent simplicial geometry as described in the following.

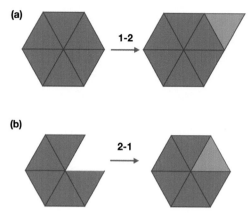

Figure 23 A schematic representation of the 2-dimensional topological moves that are allowed for the model of emergent geometry proposed in Ref. [31] with $\bar{k} = 2$. The model of emergent 2-dimensional geometry evolves in time by the subsequent addition of 2-simplices. At each time, process (a) [1-2 move] takes place with probability one, and process (b) [2-1 move] takes place with probability $p < 1$.

MODEL OF 2-DIMENSIONAL EMERGENT SIMPLICIAL GEOMETRY [31]

At time $t = 1$ the simplicial complex is formed by a single triangle. At each time $t > 1$, two processes can occur: process (a) and process (b).

(a) Let us indicate with **A** the adjacency matrix of the simplicial complex skeleton. Process (a) is defined as follows. A link $[r, s]$ having $r < s$ is chosen with probability

$$\Pi_{[r,s]} = \frac{A_{rs}(1 - \rho_{rs})}{\sum_{q<q'} A_{q,q'}(1 - \rho_{q,q'})}, \tag{5.2}$$

and a new triangle is glued to the link.

This process takes place at each time $t > 1$ with probability one.

(b) Process (b) is defined as follows. Two adjacent unsaturated links are chosen and a link connecting the two nodes at distance 2 is added to the simplicial complex together with all the triangles that this link closes as long as the move is allowed, i.e. no link is incident to more than $\bar{k}$ triangles.

This process takes place at each time $t > 1$ with probability $p < 1$.

For $\bar{k} = 2$ where the incidence number of links takes values $n_\alpha \in \{0, 1\}$ the generated simplicial complex is a 2-dimensional manifold because the evolving simplicial complex satisfies all the combinatorial conditions for having a discrete 2-dimensional manifold (defined in Section 4.1). This manifold emerges from purely combinatorial rules that make no use of its discrete geometry. Therefore this is a fundamental model of emergent geometry in $d = 2$. These manifolds have a Euler characteristic $\chi = 1$ and they describe a contractible topology; however, they display a very rich combinatorial and geometrical phenomenology. Since in the case $\bar{k} = 2$ the emergent simplicial complex is a 2-dimensional manifold (see Figure 24a), we can study the distribution of the Regge curvatures associated with the nodes, assuming that each triangle is a equilateral triangle. In this case the Regge curvature is simply related to the degree k of the nodes by

$$R_\alpha = \begin{cases} 2\pi(1 - k_{2,0}(\alpha)/6) & \text{for } \alpha \in \mathcal{B}, \\ \pi(1 - k_{2,0}(\alpha)/3) & \text{for } \alpha \in \mathcal{A}. \end{cases} \tag{5.3}$$

Since these manifolds have a Euler characteristic $\chi = 1$ according to the Gauss–Bonnet Theorem (stated in Section 4.2.1) the average Regge curvature is given by $\langle R \rangle = 2\pi$. Given Eq. (5.3) for the Regge curvature, the distribution of the Regge curvature can be derived easily from the distribution of the generalized degrees of the nodes. Since the generalized degree distribution of the nodes is exponential, the curvature distribution is also exponential. In particular, while the average curvature is positive, (i.e. $\langle R \rangle = 2\pi$) the simplicial complex can have nodes with large negative curvature in the limit $N \to \infty$.

For $\bar{k} = \infty$ and $p = 0$, where there is no limit to the number of triangles incident to each link, the network is scale-free with the same degree distribution of the Barabási–Albert model [98] (see Figure 24b). This is a notable example of emergent preferential attachment due to a topological dimension $d > 1$ of the simplicial complex. In fact, process (a) indicates that links where new triangles are attached are chosen with uniform probability. However, process (a) implies that each existing node of the network will acquire new links proportionally to how many links it is connected too, leading to an effective preferential attachment mechanism known to cause the emergence of scale-free distributions [97]. Indeed for $\bar{k} = \infty$ we have $\rho_{rs} = 0$ for every link of the network, therefore the probability $\Pi_{[r,s]}$ to attach a new triangle to a link $[r, s]$ with $r < s$ is given by

$$\Pi_{[r,s]} = \frac{A_{rs}}{\sum_{q<q'} A_{q,q'}} = \frac{A_{rs}}{L}, \tag{5.4}$$

where L indicates the total number of links in the simplicial complex. The probability $\Pi_{2,0}([r])$ that a new simplex will be incident to node r is therefore

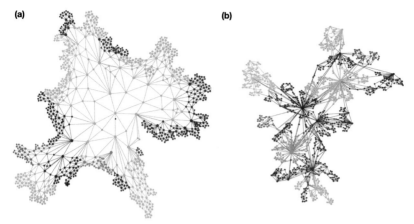

Figure 24 The network skeleton of the model of emergent 2-dimensional geometry is shown for a realization in which $\bar{k} = 2, p = 0.9$ (panel a) and for a realization in which $\bar{k} = \infty, p = 0$ (panel b).

Source: Reprinted figure from [31].

expressing the emergence of the preferential attachment mechanisms in this model. Indeed since $L = \sum_q k_q/2$ and $k_r = \sum_s A_{rs}$ we have

$$\Pi_{2,0}([r]) = \sum_{s=1}^{N} \frac{A_{rs}}{L} = 2\frac{k_r}{\sum_q k_q}, \tag{5.5}$$

expressing that at each timestep, each node r increases its degree k_r by 1 with a probability that increases linearly with its degree. This mechanism is a purely higher-order mechanism that was first pointed out in the early days of network science in Ref. [99].

5.3 Network Geometry with Flavor (Neutral Model)

5.3.1 The Definition of the NGF Model (Neutral Model)

From the previous model of emergent geometry in dimension $d = 2$ we now move to the very general framework of emergent hyperbolic geometry called Network Geometry with Flavor (NGF) [29] for emergent simplicial geometry in any topological dimension d. This is a non-equilibrium model of growing simplicial complexes in which d-dimensional simplices are subsequently glued to $(d-1)$ dimensional faces. The model depends on a parameter called *flavor* s taking values $s \in \{-1, 0, 1\}$. There are two variants of the NGF model, the neutral model and the model with fitness of the faces. Both variants of the model display notable combinatorial, topological and geometrical properties reflecting the very rich interplay between these different descriptions of the emergent hyperbolic geometry. The neutral model can be also extended to treat cell complexes, leading to additional considerations of the interplay between the topology and the geometry of NGFs. Let us start by discussing the neutral NGF simplicial complex model.

NETWORK GEOMETRY WITH FLAVOR (NEUTRAL MODEL) [29]

At time $t = 1$ the NGF is formed by a single d-dimensional simplex. At each time $t > 1$ the model evolves according to the following principles:

- GROWTH: At every timestep a new d-dimensional simplex formed by one new node and an existing $(d-1)$-face is added to the simplicial complex.
- ATTACHMENT: The probability that the new d-simplex is glued to a $(d-1)$-dimensional face α depends on the *flavor* $s \in \{-1, 0, 1\}$ and is given by

$$\Pi_\alpha^{[s]} = \frac{(1 + sn_\alpha)}{\sum_{\alpha'}(1 + sn_{\alpha'})}. \tag{5.6}$$

The role of the flavor parameter is to change the attachment probability. In particular, the attachment probability can be expressed, depending on the value of the flavor s, as

$$\Pi_\alpha^{[s]} = \frac{(1 + sn_\alpha)}{\sum_{\alpha'}(1 + sn_{\alpha'})} \propto \begin{cases} 1 - n_\alpha & \text{if} \quad s = -1, \\ \text{const} & \text{if} \quad s = 0, \\ k_{d,d-1}(\alpha) & \text{if} \quad s = 1. \end{cases} \tag{5.7}$$

Therefore depending on the value of the flavor the NGF implements different combinatorial rules.

- For $s = -1$ the attachment probability $\Pi_\alpha^{[-1]}$ is non-zero if $n_\alpha = 0$ but becomes zero as soon as $n_\alpha = 1$. Therefore we obtain simplicial complexes that are discrete manifolds (defined in Section 4.1) in which the incidence numbers can only take values $n_\alpha \in \{0, 1\}$.
- For $s = 0$ the attachment probability $\Pi_\alpha^{[0]}$ is uniform over all the (d-1)-dimensional faces. Therefore the simplicial complexes generated by the model have incidence numbers taking any non-negative integer value $n_\alpha \geq 0$.
- For $s = 1$ the attachment probability $\Pi_\alpha^{[1]}$ is proportional to the generalized degree $k_{d,d-1}(\alpha)$ of the $(d-1)$-dimensional face α. Therefore the probability of attaching a d-simplex to a face α increases with the number of

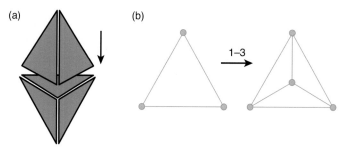

Figure 25 The NGF model in $d = 3$ dimensions evolves in time by the subsequent addition of 3-simplices (tethrahedra) to 2-dimensional faces (triangles). For $s = -1$ every $(d-1)$-dimensional face can be incident to at most two d-dimensional simplices. Here we schematically show a single topological move for $d = 3$, $s = -1$ (panel a) and its planar projection on the plane formed by the $(d-1)$-dimensional face (triangle) (panel (b)). In this planar projection, the attachment of the new tetrahedron to the initial triangle induces a triangulation of the initial triangle into three distinct triangles. For this reason this topological move is also called a $1-3$ topological move.

d-faces already attached to it, generalizing the preferential attachment mechanisms of growing scale-free network models [98]. In this case the simplicial complexes generated by the model have incidence numbers taking any nonnegative integer value $n_\alpha \geq 0$. Note that the NGF with $d = 1$ and $s = 1$ reduces to the Barabási–Albert model [97].

In Figure 25 we provide a visual representation of the basic NGF process for $d = 3$ and $s = -1$, i.e. the gluing of a tetrahedron to a triangular face. This move is one of the fundamental topological moves called the $1 - 3$ Pachner move (see Appendix D for more details on topological moves). It is interesting to notice that if we consider the projection of the new added tetrahedron on the plane formed by the triangle on which the tetrahedron is glued, the $1-3$ Pachner move corresponds to a triangulation of the original triangle. Thus for $d = 3$ and $s = -1$ the network skeleton of the NGF reduces to a random Apollonian network (see Appendix G). For general dimensions d the NGF evolves through the iterations of $1 - d$ Pachner moves. Despite the simplicity of the model, generating contractible topology whose unique non-zero Betti number is $\beta_0 = 1$ (with Euler characteristic $\chi = 1$) the NGFs display a very rich structure characterized by a profound interplay between their combinatorial, topological and geometrical properties.

5.3.2 Combinatorial Properties of NGFs

The combinatorial properties of NGFs show a strong dependence on the NGF topology. Indeed the m-dimensional faces of the same NGF can have very different statistical properties for different values of m. In particular, the generalized degree distributions depend not only on the dimension d of the simplicial NGF but also on the dimension m of the faces considered. The emergence of different statistical properties of faces of different dimension m belonging to the same simplicial complex is an effect due to the higher-order nature of the NGF that does not have any equivalent in network models. This higher dimensionality of NGFs is also responsible for the emergence of preferential attachment in the NGF model even if the NGF dynamics do not include any specific preferential attachment rule, i.e. the flavor has values $s = 0$ (uniform attachment) or $s = -1$ (generating discrete manifolds). This phenomenon is apparent if we consider the probability $\Pi_{d,m}^{[s]}(k)$ of attaching a d-dimensional simplex to an m-dimensional face in the NGF evolution [29]. This is given by (see Appendix E for details)

$$
\Pi_{d,m}^{[s]}(k) = \begin{cases} \frac{2-k}{(d-1)t} & \text{for} \quad d - m + s = 0, \\ \frac{1-s}{(d+s)t} & \text{for} \quad d - m + s = 1, \\ \frac{(d-m-1+s)k+1-s}{(d+s)t} & \text{for} \quad d - m + s > 1. \end{cases} \tag{5.8}
$$

Therefore for $d - m + s > 1$ the probability $\Pi_{d,m}(k)$ grows linearly with the generalized degree of the face and we observe a generalized preferential attachment as a consequence of the geometry and dimensionality of the NGF. This implies that the generalized degree of the nodes ($m = 0$) is power-law distributed as long as $d \geq 2-s$. For $s = 0$ and $d = 2$, if we attach triangles to random links chosen with uniform probability, we actually end up attaching new links to nodes chosen with a probability that grows linearly with their degree (preferential attachment) according to the mechanisms explained in Section 5.2. For $s = -1$ and $d = 3$, an intuition of the emergent preferential attachment of the NGF model can be provided by Figure 26: as a node increases its generalized degrees, the number of faces with incidence number $n_\alpha = 0$ to which we can glue a new d-dimensional simplex increases, leading to a generalized preferential attachment. Using Eq. (5.8) it is straightforward to calculate the distribution $P_{d,m}^{[s]}(k)$ of the generalized degrees of the m-dimensional faces for any value of d and s. These generalized degree distributions follow a regular pattern depending on d, m and s (see Table 1). The details of this derivation are provided in Appendix F. In particular, the generalized degree distribution is bimodal for $d-m+s = 0$, exponential for $d-m+s = 1$ and power-law for $d-m+s > 1$. Therefore if we consider an NGF with $d = 3$ and $s = -1$, the triangles (2-dimensional faces), links (1-dimensional faces) and nodes (0-dimensional faces) have bimodal, exponential and scale-free generalized degree distributions, respectively. From these results we can draw two important conclusions regarding the statistical properties of emergent manifolds:

Table 1 Distribution of generalized degrees of faces of dimension m in a d-dimensional NGF of flavor s at $\beta = 0$. For $d \geq 2m + 2 - s$ the power-law distributions are scale-free, i.e. the second moment of the distribution diverges.

Flavor	$s = -1$	$s = 0$	$s = 1$
$m = d - 1$	Bimodal	Exponential	Power-law
$m = d - 2$	Exponential	Power-law	Power-law
$m \leq d - 3$	Power-law	Power-law	Power-law

(a)   **(b)**

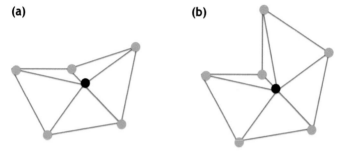

Figure 26 A schematic representation of the geometrical emergence of the preferential attachment for the NGF model with $d = 3, s = -1$. We consider a node (the black node) incident to 3 tetrahedra and to 5 triangles with incidence number $n_\alpha = 0$ (panel (a)). When a new tetrahedron incident to the black node is added to the simplicial complex (panel (b)), the number of triangles incident to the black node and with incidence number $n_\alpha = 0$ increases to 6, indicating that the probability of attaching a new d-simplex incident to the black node increases as we increase its generalized degree.

- Provided the dimension d is sufficently high, i.e. $d \geq 3$, the generation of scale-free manifolds is the natural outcome of the NGF neutral model. This shows that the complexity of these structures arises from the very simple and therefore fundamental rules for emergent geometry captured by the NGF model with $s = -1$ [39].

- Despite the very strong power-law fluctuations observed in the generalized degree of faces of dimension $m \leq d - 2$, the NGF manifolds are characterized by a more moderate exponential fluctuation of the generalized degree of faces of dimension $m = d - 2$. This is interesting in light of the fact that the generalized degrees of faces of dimension $m = d - 2$ are related to the Regge curvature according to Eq. (4.3) in the hypothesis in which all the simplices of the NGF are identical. Therefore, even in dimensions $d \geq 3$, the distribution of the Regge curvature of the NGF manifolds always remains exponential, as long as we assume that the simplices are all identical.

In the NGF, the degree k_r of a node r is related to its generalized degree $k_{d,0}([r])$ by the simple relation $k_r = k_{d,0}([r]) + (d-1)$. Therefore it is straightforward to derive the degree distribution of the NGF from the generalized degree distribution of the nodes. Excluding the case $(d,s) = (1,-1)$ for which the NGF is a linear chain, the degree distribution $P_d^{[s]}$ of an NGF of dimension d and flavor s is given by the following expressions:

$$P_d^{[s]}(k) = \left(\frac{d}{d+1}\right)^{k-d}\frac{1}{d+1}, \tag{5.9}$$

for $d + s = 1$; and for $d + s > 1$ it is given by

$$P_d^{[s]}(k) = \frac{\Gamma[k-d+d/(d+s-1)]}{\Gamma[k-d+1+(2d+s)/(d+s-1)]}, \tag{5.10}$$

with

$$C = \frac{d+s}{2d+s}\frac{\Gamma[1+(2d+s)/(d+s-1))]}{\Gamma[d/(d+s-1)]}. \tag{5.11}$$

This last expression of the degree distribution for large values of the degree k, i.e. for $k \gg 1$, can be approximated by a power-law distribution

$$P_d^{[s]}(k) = Ck^{-\gamma_d^{[s]}}, \tag{5.12}$$

with power-law exponent

$$\gamma_d^{[s]} = 2 + \frac{1}{d+s-1}. \tag{5.13}$$

The following considerations are therefore in place:

- NGFs with $s = 1$ implementing a generalized preferential attachment are always scale-free;
- NGFs with $s = 0$ implementing a uniform attachment are scale-free for $d \geq 2$;
- NGFs with $s = -1$ which generate discrete manifolds are scale-free for $d \geq 3$.

Therefore, as long as the dimension of the simplicial complex is sufficiently large, the NGF skeleton is a scale-free network, regardless of the value of the flavor. If follows that an explicit preferential attachment is not necessary to get a

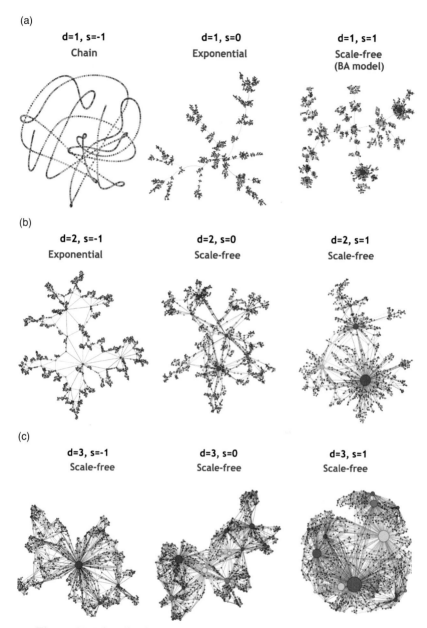

(a)

d=1, s=-1
Chain

d=1, s=0
Exponential

d=1, s=1
Scale-free
(BA model)

(b)

d=2, s=-1
Exponential

d=2, s=0
Scale-free

d=2, s=1
Scale-free

(c)

d=3, s=-1
Scale-free

d=3, s=0
Scale-free

d=3, s=1
Scale-free

Figure 27 Visualization of the NGF network skeletons for dimension
$d \in \{1, 2, 3\}$ and flavor $s \in \{-1, 0, 1\}$.

Source: Reprinted figure with permission from [29] ©Copyright (2016) by the American Physical Society.

scale-free network skeleton, provided the dimension of the simplicial complex is large enough. The skeleton of an NGF of dimension $d \in \{1, 2, 3\}$ is shown in Figure 27.

Figure 28 The emergent hyperbolic geometry of the NGF with $d = 2$ and $s = -1$ is revealed by the fact that this network, generated exclusively by combinatorial rules generates a random Farey tree with natural embedding in the Poincaré disk.

5.3.3 Emergent Hyperbolic Geometry

Assuming that all the links of NGF simplicial complexes have equal length, it is not possible to embed NGFs in any Euclidean space. Network Geometries with Flavor are indeed simplicial complexes with a small-world network skeleton, i.e. an infinite Hausdorff dimension $d_H = \infty$. A related feature of NGFs related to their small-world nature is that the boundaries $\mathcal{A}$ of NGFs with flavor $s = -1$ have an area A that grows linearly with the number of nodes, i.e. $A \propto N$. These observations are already a strong hint that NGFs are hyperbolic. A closer look reveals that NGFs satisfy the Gromov criteria for δ-hyperbolicity (see Section 4.2.2) for every flavor s. In fact they are δ-hyperbolic with $\delta = 1$, for every flavor s and dimension d. The hyperbolic nature of NGFs is even more apparent for the case of NGFs of flavor $s = -1$, which define hyperbolic discrete manifolds [30]. In particular the NGF with dimension $d = 2$ and flavor $s = -1$ generates simplicial complexes whose skeletons are random Farey graphs (see Figure 28). Farey graphs are $d = 2$ hyperbolic lattices which tesselate the Poincaré disk having all *ideal* nodes, i.e. all nodes at the boundary of the Poincaré disk. For $d = 3$ and $s = -1$ the NGFs are hyperbolic manifolds having all ideal nodes, i.e. all the nodes are at the boundary of the 3-dimensional hyperbolic space.

These observations reveal that the NGF is a model of emergent hyperbolic network geometry, i.e. it is a model that evolves according to abstract

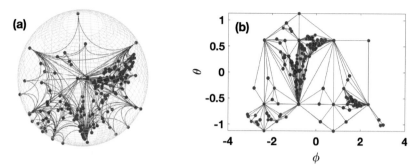

Figure 29 The NGF of dimension $d = 3$ and flavor $s = -1$ describes an hyperbolic manifold that tessellates the hyperbolic space $\mathbb{H}^3$ having all ideal nodes, i.e. all the nodes are at the boundary of the hyperbolic space (panel (a)). If we allow the links of the NGF with dimension $d = 3$ and flavor $s = -1$ to have different lengths, this same embedding can be interpreted as embedding into a Euclidean ball. By projecting the network on the surface of this ball, we reveal that the NGF with dimension $d = 3$ and flavor $s = -1$ has a network skeleton which is planar and equivalent to a random Apollonian network (panel (b)).

Source: Reprinted from Ref. [32].

combinatorial rules but it is able to generate simplicial complexes with notable geometrical properties.

Interestingly, if we allow the links of the NGF to have different lengths, we can interpret the hyperbolic embedding of the NGF shown in Figures 28 and 29(a) as an embedding in a 3-dimensional Euclidean ball. Finally by projecting the NGF skeleton on the surface of this ball we reveal that this "holographic" projection of the NGF has a network skeleton which is planar and reduces (see Appendix G for details) to a random Apollonian graph (see Figure 29(b)) which has notable geometrical properties (see Appendix G). In particular, the skeleton of a d-dimensional NGF with boundary can be naturally interpreted as the skeleton of a $(d-1)$-dimensional simplicial complex without boundary, as long as $d \geq 2$.

The NGF are models of emergent geometry, and as such their combination makes no use of their underlying geometry. However, from their dynamics it emerges that they can be considered as a models of growing connected percolation clusters on regular hyperbolic simplicial complexes; in particular the NGF with flavor $s = -1$ can be considered as a model of a growing (simplicial complex) percolation cluster on the Apollonian simplicial complex, while the NGF with flavor $s = 0$ and flavor $s = 1$ can be considered as a model of growing (simplicial complex) percolation clusters on the pseudo-fractal simplicial complex

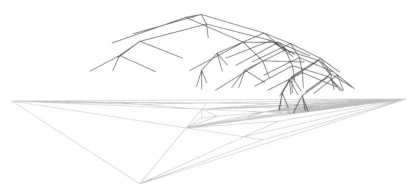

Figure 30 The network skeleton of the NGF of dimension $d = 3$ and $s = -1$ is schematically represented on a 2-dimensional plane (random Apollonian graph). The NGF is related to a tree whose nodes correspond to the tetrahedra of the NGF and whose links correspond to the triangles shared by two tetrahedra.

(for a discussion of Apollonian and pseudo-fractal simplicial complexes see Appendix G).

We conclude this section by observing that the NGF of flavor $s = -1$ has an evolution that can be mapped to one of the trees whose nodes correspond to the d-simplices of the NGF and whose links correspond to $(d - 1)$-faces connecting two d-simplices, i.e. with incidence number $n_\alpha = 1$ (see Figure 30).

5.3.4 Emergent Spectral Dimension of NGFs

Network Geometries with Flavor have a network skeleton that is small-world, i.e. they have infinite Hausdorff dimension $d_H = \infty$. However, the relevant geometrical nature is revealed by the fact that they display a finite spectral dimension d_S. Indeed the cumulative density of eigenvalues $\rho_c(\lambda)$ scales according to a power-law [Eq. (4.12)] with exponent $d_S/2$.

The scientific interest in finite spectral dimension and its effect on diffusion dynamics is increasing as this is likely to be a rather neglected aspect of real networks and not as rare as previously assumed [31]. We mention here that not only the NGF but also the $d = 2$ model for emergent geometry described in Section 5.2 and several other models have been shown to display finite spectral dimensions [31, 101–103].

As a side note, we want to point out that in a recent paper [100] it has been shown that NGF allows an interpretation as a network of quantum oscillators. In this interpretation the spectral dimension parametrizes the density

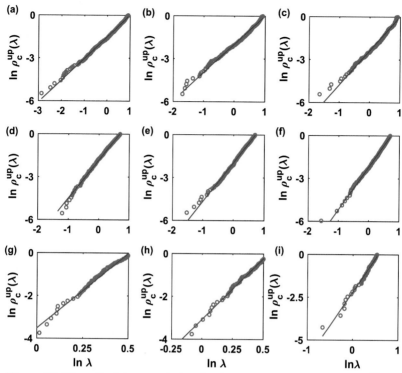

Figure 31 The tail of the cumulative density of non-zero eigenvalues $\rho_c^{up}(\lambda)$ of the m-order up-Laplacians is shown in a log-log plot (data points) for the NGF (simplicial complex) of dimension $d = 3$ and flavor $s = -1$ (panels a, d, g for $m = 0, 1, 2$ respectively), $s = 0$ (panels b, e and g for $m = 0, 1, 2$, respectively) and $s = 1$ (panels c, f and i for $m = 0, 1, 2$, respectively) together with its fit.

Source: Reprinted from [33].

of states of the oscillators and can be inferred efficiently by a quantum probe.

The NGF simplicial complexes also display higher-order spectral dimensions. Indeed, the same NGF simplicial complex can be characterized by a set of spectral dimensions $d_S^{[m]}$ characterizing the spectral dimensions of the m-dimensional up-Laplacians with $0 \le m \le d-1$ (see Figure 31) [33]. The values of these spectral dimensions depend on m, indicating that the the higher-order diffusion processes from m simplices to m simplices through the $(m + 1)$-dimensional simplices have different statistics for the return-time probability distributions. The higher-order spectral dimensions are also a feature of the deterministic version of NGF, including Apollonian and pseudo-fractal simplicial complexes, and can be predicted using the renormalization group [88].

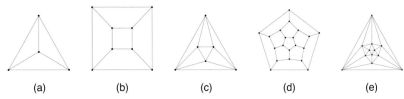

Figure 32 The planar graphs that describe the skeleton of the regular polytopes in $d = 3$ (Platonic solids): (a) tetrahedron, (b) cube, (c) octahedron, (d) dodecahedron, (e) icosahedron.

Source: Reprinted from [32]

5.4 Generalization of NGF to Cell Complexes

5.4.1 Combinatorial Properties of the NGF Cell Complexes

The NGF simplicial complex model can be generalized to treat NGF cell complexes [32]. The NGF cell-complex model has the same definition as the NGF simplicial complex, the only difference being that any d-simplex is substituted with a regular d-dimensional polytope. In Figure 32 we show the planar representation of the skeleton of the Platonic solids (regular polytopes in dimension $d = 3$). The NGF cell complex in dimension $d = 3$ will start from a single Platonic solid, let us say a cube, and at each timestep will glue a cube to a single $(d-1)$-dimensional face α (square) of the cell complex, according to the probability $\Pi_\alpha^{[s]}$ given by Eq. (5.6). While for NGF with $s = -1$ the addition of a new simplex reduces to a $1 - d$ Pachner move, more general topological moves apply if we consider cell complexes. For instance in $d = 3$ and $s = -1$, while the $1 - 3$ Pachner move for a simplicial complex can be seen as a substitution of one triangle with three triangles at the boundary of the simplicial complex, if we work, let us say, with cubes, we substitute a square with five squares at the boundary of the cell complex (see Figure 32). Therefore changing the polytope, i.e. the building block of the NGF, changes the area/volume (i.e. A/N) ratio of the manifold [104].

The combinatorial properties of NGF cell complexes deserve special attention. Like NGF simplicial complexes, the degree distribution of the 1-skeleton of the NGF cell complex is exponential for $d = 1, s = 0$ and for $d = 2, s = -1$. For all other values of the dimension d and the flavor s it is power-law. However, the power-law exponents of NGF cell complexes given in Table 2 deliver some surprises. Indeed, these exponents are not always in the range $\gamma \in (2, 3]$ indicating the scale-free universality class.

From Table 2 we can draw three main, unexpected conclusions which reveal additional non-trivial effects of geometry on the statistical properties of higher-order networks:

Table 2 Power-law exponent γ of the degree distribution of Network Geometry with Flavor s built by gluing regular, convex polytopes in dimension d.

Flavor/Cell	$s = -1$	$s = 0$	$s = 1$
d = 1			
link	N/A	N/A	3
d = 2			
p-polygon	N/A	p	$1 + \frac{p}{2}$
d = 3			
tetrahedron	3	$2\frac{1}{2}$	$2\frac{1}{3}$
cube	5	$3\frac{1}{2}$	3
octahedron	4	$3\frac{1}{3}$	3
dodecahedron	11	$6\frac{1}{2}$	5
icosahedron	7	$5\frac{3}{4}$	5
d = 4			
pentachoron	$2\frac{1}{2}$	$2\frac{1}{3}$	$2\frac{1}{4}$
tesseract	4	$3\frac{1}{3}$	3
hexadecachoron	$3\frac{1}{3}$	$3\frac{1}{7}$	3
24-cell	$6\frac{1}{2}$	$5\frac{3}{5}$	5
120-cell	60	$40\frac{2}{3}$	31
600-cell	$34\frac{2}{9}$	$32\frac{10}{19}$	31
d > 4			
simplex	$2 + \frac{1}{d-2}$	$2 + \frac{1}{d-1}$	$2 + \frac{1}{d}$
cube	$3 + \frac{2}{d-2}$	$3 + \frac{1}{d-1}$	3
orthoplex	$3 + \frac{1}{2^{(d-2)}-1}$	$3 + \frac{1}{2^{d-1}-1}$	3

Source: Reprinted table from [32].

- Simplicial complexes with power-law degree distribution are always scale-free, i.e. they have a power-law exponent $\gamma \in (2,3]$ and diverging second moment of the degree distribution.
- Other cell complexes are scale-free only if they have flavor $s = 1$ (preferential attachment).
- Some cell complexes in $d = 2, 3, 4$ are not even scale-free for flavor $s = 1$.

Therefore, if NGFs have simplices as their buidling blocks, scale-free network skeletons can be obtained for every flavor s as long as the dimension d is large enough, i.e. $d \geq 2 - s$. However, if the building blocks of the NGF are not simplices it is much more difficult to obtain scale-free network skeletons. Indeed, in this case having a generalized preferential attachment, i.e. a

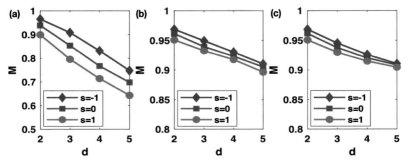

Figure 33 The average modularity M of NGFs of different flavor s is plotted versus their dimension in the case of simplicial complexes (panel (a)), and of cell complexes formed by hypercubes (panel (b)) or orthoplexes (panel (c)).

Source: Reprinted from Ref. [32].

flavor $s = 1$, is a necessary but not sufficient condition to obtain scale-free network skeletons.

5.4.2 Emergent Community Structure of NGFs

Network Geometries with Flavor generate network skeletons which display emergent communities. In fact, NGFs are related to the model with triadic closure treated in Section 3.5 that also displays emergent communities. The emergent community structure is a property of NGFs of any dimension d and flavor s and extends not only to the original (neutral) definition of NGF simplicial complexes but also to NGF cell complexes. In Figure 33 the average modularity M (proposed in Ref. [105] to measure the significance of the community structure) of NGFs is shown as a function of the dimension of the simplicial or cell complex. The modularity M is shown to decrease with dimension d but keeps high values for any dimension $d \in \{2, 3, 4, 5\}$.

5.4.3 Spectral Dimension of NGFs Remains as in Cell Complexes

We have shown in the previous section that the NGF simplicial complexes have a geometrical nature which is responsible for the fact that NGFs display finite spectral dimensions of the graph Laplacian and of the higher-order up-Laplacians as well. In particular we have shown that NGFs are characterized by a vector of spectral dimensions indicating the statistical properties of diffusion and higher-order diffusion occurring in the same simplicial complex. Here we want to explore how the spectral dimension of the graph Laplacian changes if we consider NGF cell complexes having different dimension d when they are formed by different polytopes (i.e. different building blocks). Interestingly, the spectral dimension d_S found numerically by fitting the cumulative density of

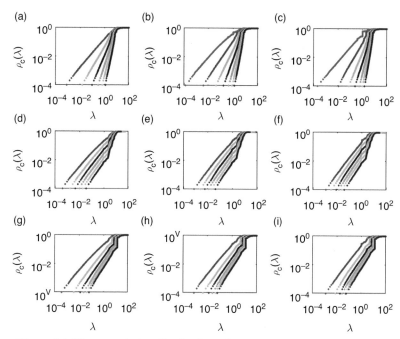

Figure 34 The cumulative distribution of eigenvalues $\rho_c(\lambda)$ of the graph Laplacian is shown versus λ for the Network Geometry with Flavor formed by polytopes of dimensions $d \in \{1, 2, 3, 4, 5, 6\}$. Panels (a)-(d)-(g), (b)-(e)-(h), (c)-(f)-(i) refer to NGFs with flavor $s = -1$, $s = 0$ and $s = 1$, respectively. Panels (a)-(b)-(c), (d)- (e),-(f) and (g)-(h)-(i) refer to NGFs formed by d-dimensional simplices, hypercubes and orthoplexes, respectively.

Source: Reprinted from [32].

eigenvalues increases superlinearly with the dimension d for NGF simplicial complexes, while it appears to saturate at high dimensions d for cell complexes (see Figures 34 and 35). Therefore the spectral dimension of NGFs can change if the the cell (building block) by which the NGF is constructed changes. As the nature of the building block used to construct the cell complexes changes the local topological move of the NGF and the area/volume (i.e. A/N) ratio of the cell complex, the fact that the spectral dimension of the NGF depends on the polytope used for the evolution of the NGF is again another important effect that topology has on the geometrical aspect of these higher-order networks [104].

5.4.4 Other Generalizations of NGFs

In Ref. [106] the NGF simplicial complex model is generalized by considering two possible variations of the original simplicial NGF model (see Figure 36).

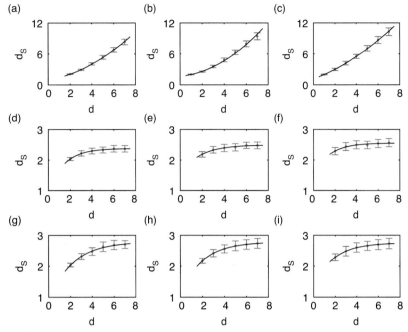

Figure 35 The spectral dimension d_S of the graph Laplacian is shown versus d for the Network Geometries with Flavor formed by polytopes of dimensions $d \in \{1, 2, 3, 4, 5, 6\}$. Panels (a)-(d)-(g), (b)-(e)-(h), (c)-(f)-(i) refer to NGFs with flavor $s = -1$, $s = 0$ and $s = 1$, respectively. Panels (a)-(b)-(c), (d)-(e)-(f) and (g)-(h)-(i) refer to NGFs formed by d-dimensional simplices, hypercubes and orthoplexes, respectively.

Source: Reprinted from [32].

The first variation entails allowing any new node to connect to the existing simplicial complex by gluing $\bar{k} \geq 1$ simplices to $\bar{k}$ existing $(d-1)$-faces. The second variation entails considering weighted simplices, so that every d-dimensional simplex is associated with a weight, and reinforcing the weights of the d-simplices in time. The first variation does not change the large-scale properties of the degree distribution and leaves unchanged the power-law exponent of any power-law degree distribution. However, it changes the geometric properties of the NGF as long as $\bar{k} > 1$. In particular, for $\bar{k} > 1$ the NGF network skeleton becomes mean-field and develops a finite spectral gap. The second variation of the NGF model, reinforcing the weight of the simplices in time, also changes the statistical properties of NGFs and can result in highly nontrivial correlations between the weights of the simplices and their generalized degrees. In Ref. [107] a further variation of the simplicial NGF model is considered, allowing new d-dimensional simplices to be attached to a face α chosen

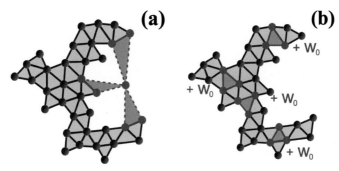

Figure 36 Schematic representation of the two variations of the original NGF model considered in Ref. [104]. Panel (a): a new node is allowed to connect to the existing simplicial complex by more than one d-simplex. Panel (b): d-dimensional simplices can be weighted and their weight can be reinforced dynamically in time.

Source: Reprinted figure with permission from [104] ©Copyright (2017) by the American Physical Society.

according to uniform attachment, (i.e. according to $\Pi_\alpha^{[0]}$) with probability q and according to generalized preferential attachment (i.e. according to $\Pi_\alpha^{[1]}$) with probability $1 - q$.

5.5 Network Geometry with Flavor (with Fitness)

5.5.1 Introducing the Energy of the Faces

At each timestep of the evolution of the neutral NGF model, the future non-equilibrium dynamics of the simplicial complex only depends on its present topological properties, i.e. on the incidence number of its $(d - 1)$-dimensional faces. However the non-equilibrium dynamics of simplicial complexes, and of networks as well, can also depend on some non-topological variables. By considering a parallelism to quantum gravity, the introduction of non-topological variables associated with faces of a simplicial complex determining the simplicial complex dynamics can be seen as the complex system equivalent of introducing mass terms in quantum gravity. If we focus on complex system network models, with $d = 1$, the non-topological variable associated with a node can combine information coming from available meta-data describing the node properties and can modulate the ability of a node to acquire new links, also called the node fitness. The Bianconi–Barabási model [95, 106] is the reference network model that describes the evolution of networks where nodes are associated with a fitness value and has been used to model a variety of real networks, including the World Wide Web. In this context the fitness indicates some quantitative measure of the perceived quality of the content of a webpage (i.e. a node in the network).

In this section we will address the Network Geometry of Flavor (NGF) formed by d-dimensional simplicial complexes whose faces are associated with an intrinsic property called *energy* which describes the non-topological features associated with them [29]. From the energy of a face one can determine its *fitness*, which describes the rate at which the face increases its generalized degree. The NGF model with fitness of the faces generalizes the Bianconi–Barabási model by associating with each m-dimensional face an energy and a fitness.

ENERGY AND FITNESS OF THE FACES OF NGF SIMPLICIAL COMPLEXES [29]

The energy ε_α of the m-dimensional face α indicates its intrinsic (non-topological) properties. The energy $\varepsilon_{[r]}$ of a node r is a non-negative number drawn from a given distribution $g(\varepsilon)$. The energy of a face α of dimension $m > 0$ is the sum of the energies of the nodes belonging to it, i.e.

$$\varepsilon_\alpha = \sum_{r \subset \alpha} \varepsilon_{[r]}. \qquad (5.14)$$

The *fitness* associated with an m-dimensional face α describes the rate at which the face increases its generalized degree and is given by

$$\eta_\alpha = e^{-\beta \varepsilon_\alpha} \qquad (5.15)$$

where $\beta > 0$ is a parameter called *inverse temperature*. For $\beta = 0$ all the fitness are the same, and equal to one, while for $\beta \gg 1$ the small difference in energy leads to big differences in the fitnesses of the faces.

Figure 37 describes schematically how energies are assigned to higher-dimensional faces of the NGF simplicial complex. The NGF evolution can be modified to take into consideration the effect of assigning different fitness to the NGF faces.

NETWORK GEOMETRY WITH FLAVOR (WITH FITNESS) [29]

At time $t = 1$ the simplicial complex is formed by a single d-dimensional simplex. Each node r of this simplex has energy $\varepsilon_{[r]}$ drawn from a $g(\varepsilon)$ distribution. The energies of the higher-dimensional faces are calculated according to Eq. (5.14).

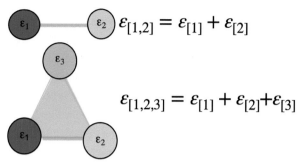

$$\varepsilon_{[1,2]} = \varepsilon_{[1]} + \varepsilon_{[2]}$$

$$\varepsilon_{[1,2,3]} = \varepsilon_{[1]} + \varepsilon_{[2]} + \varepsilon_{[3]}$$

Figure 37 Schematic representation indicating how the energies associated with the links and triangles of an NGF are calculated starting from the node energies.

- GROWTH: At every timestep a new d-dimensional simplex formed by one new node and an existing $(d-1)$-face is added to the simplicial complex. Each new node r has energy $\varepsilon_{[r]}$ drawn from a $g(\varepsilon)$ distribution. The energies of the new higher-dimensional faces are calculated according to Eq. (5.14).
- ATTACHMENT: At every timestep the probability that the new d-simplex is connected to the existing $(d-1)$-dimensional face α depends on the *flavor* $s \in \{-1, 0, 1\}$ and on the *inverse temperature* $\beta > 0$ and is given by

$$\Pi_\alpha^{[s]} = \frac{e^{-\beta\varepsilon_\alpha}(1 + sn_\alpha)}{\sum_{\alpha'} e^{-\beta\varepsilon_{\alpha'}}(1 + sn_{\alpha'})}. \tag{5.16}$$

For $\beta = 0$ the NGF (with fitness of the m-faces) reduces to the neutral NGF model, i.e. $\Pi_\alpha^{[s]}$ reduces to Eq. (5.6).

The role of the inverse temperature $\beta > 0$ is to bias the evolution of the simplicial complex in such a way that faces with lower energy increase their generalized degree faster. This model for $d = 1$ and $s = 1$ reduces to the Bianconi–Barabási model [96, 108] which displays emergent Bose–Einstein statistics and the Bose–Einstein condensation of complex networks. The Bose–Einstein condensation observed in this model is a topological phase transition in which the network is dominated by a succession of super-hub nodes, i.e. nodes with a degree growing linearly with time, with at most a logarithmic correction. In the next paragraphs we will discuss how this scenario changes for NGFs. We will discover notable statistical and topological properties of NGFs with fitness of the faces and inverse temperature $\beta > 0$. In particular, we will show that not only Bose–Einstein statistics but also Fermi–Dirac statistics describe the statistical properties of the NGF faces and we will reveal that in the same NGF the statistics followed by faces of different dimensions can vary. Moreover, we will discuss the relation between the total energy of the NGF manifolds with flavor $s = -1$ describing the probability of a given causal evolution of the NGF, and the Regge curvature. Finally, we will discuss the properties of the topological phase transitions that occur for the NGF in the low-temperature regime and extend to the simplicial complex realm the Bose–Einstein condensation in complex networks.

5.5.2 Probability of an NGF Evolution

Each NGF is the result of an evolution caused by the subsequent addition of d-simplices, where the addition of any new simplex to a face α has probability given by Eq. (5.16). Therefore the probability of a given NGF evolution of size N is given by the product of the probabilities of each single addition of a new simplex up to time $t = N$. Here we consider in particular the probability of the evolution leading to NGFs and we show that for $s = -1$ this probability can be related to the Regge curvature of the $(d-2)$-faces of the simplicial complex (defined in Section 4.2.1).

ENERGY OF NGFs

An NGF is a simplicial complex $\mathcal{K}$ generated by the subsequent addition of N d-simplices attached to existing $(d-1)$-faces. The history $\mathcal{I}$ of an NGF evolution up to time $t = N$ is formed by the list of faces $\alpha(t)$ chosen by the NGF algorithm up to time t, i.e. $\mathcal{I} = \{\alpha(2), \dots \alpha(N)\}$. As long as the generalized degree of the nodes reaches a steady state, the history $\mathcal{I}$ of an NGF evolution has probability

$$P(\mathcal{I}) = \frac{1}{N!} e^{-\beta(E - \mu_{d,d-1}^{[s]} N) + \hat{\Sigma}} \tag{5.17}$$

where E *the total energy of the NGF* is given by

$$E = \sum_{\alpha \in \mathcal{S}_{d-1}(\mathcal{M})} \varepsilon_\alpha n_\alpha, \tag{5.18}$$

$\hat{\Sigma}$ is given by

$$\hat{\Sigma} = \begin{cases} 0 & \text{for} \quad s \in \{-1, 0\}, \\ \sum_{\alpha \in \mathcal{S}_d(\mathcal{K})} \ln[(1 + s n_\alpha)!] & \text{for} \quad s = 1, \end{cases} \tag{5.19}$$

and the *chemical potential* $\mu_{d,d-1}^{[s]}$ is defined as

$$\lim_{N\to\infty} \frac{\sum_{\alpha\in\mathcal{S}_{d-1}(\mathcal{K})} e^{-\beta\varepsilon_\alpha}(1+sn_\alpha)}{N} = e^{-\beta\mu_{d,d-1}^{[s]}}. \tag{5.20}$$

Here we consider the relation between the total energy of the NGF manifold with flavor $s = -1$ and the Regge curvature. Assuming that all the d-simplices are identical and have dihedral angle θ_d, the Regge curvature is given by Eq. (4.3). By considering the combinatorial properties of the generalized degrees of simplicial complexes (i.e. Eq. (2.3)), it can be shown [29] that the total energy E of the NGF manifolds (with flavor $s = -1$) can be expressed in terms of the Regge curvatures associated with its $(d-2)$-faces as

$$E = \frac{B_d}{\theta_d}\left(\Lambda - \sum_{\alpha\in\mathcal{S}_{d-2}(\mathcal{M})} \varepsilon_\alpha R_\alpha\right), \tag{5.21}$$

where $B_d = 2/(d-1)$ and

$$\Lambda = \left(\pi - \frac{\theta_d}{2}\right)\sum_{\alpha\in\mathcal{S}_{d-2}(\mathcal{M})} \varepsilon_\alpha. \tag{5.22}$$

Since every discrete manifold $\mathcal{M}$ can result from distinct histories $\mathcal{I}$, in Ref. [29] the probability of observing a given discrete manifold $\mathcal{M}$ can be calculated by summing over all the possible histories leading to the same manifold $\mathcal{M}$.

ENTROPY OF NGF MANIFOLDS

As long as the generalized degree of the nodes reaches a steady state, the probability $P(\mathcal{M})$ that an NGF with flavor $s = -1$ generates a given manifold $\mathcal{M}$, independent of its history $\mathcal{I}$ is given by [29]

$$P(\mathcal{M}) = e^{-\beta E - v_{d,d-1}N}, \tag{5.23}$$

where E is the total energy of the manifold given by Eq. (5.18) or equivalently Eq. (5.21) and $v_{d,d-1}$ is a constant. The *entropy* $\tilde{S}$ of NGF manifolds (with $s = -1$) evolved up to time $t = N$, defined as

$$\tilde{S} = -\sum_{\mathcal{M}} P(\mathcal{M}) \ln P(\mathcal{M}) \tag{5.24}$$

can be expressed as

$$\tilde{S} = \beta \langle E\rangle + v_{d,d-1}N, \tag{5.25}$$

where the expectation of the total energy $\langle E \rangle$ is calculated over the distribution $P(\mathcal{M})$. It follows that $\tilde{S}$ is extensive as the area A of the NGF, i.e. $\tilde{S} \propto N$ and $A \propto N$. This result implies that for NGF with flavor $s = -1$, the entropy $\tilde{S}$ is proportional to the area A,

$$\tilde{S} \propto A. \tag{5.26}$$

5.5.3 Emergent of Quantum Statistics

The NGFs display very special combinatorial properties which reflect the rich interplay between the energies associated with their faces and their expected generalized degrees. Indeed the statistical properties of the m-dimensional faces are captured by the Fermi–Dirac $n_F(\varepsilon, \mu)$, the Bose–Einstein $n_B(\varepsilon, \mu)$ and the Boltzman $n_Z(\varepsilon, \mu)$ distributions, defined as

$$n_F(\varepsilon, \mu) = \frac{1}{e^{\beta(\varepsilon-\mu)} - 1},$$
$$n_Z(\varepsilon, \mu) = e^{-\beta(\varepsilon-\mu)},$$
$$n_B(\varepsilon, \mu) = \frac{1}{e^{\beta(\varepsilon-\mu)} + 1}. \tag{5.27}$$

These statistics are known in physics to represent the occupation of energy levels in a gas of quantum Fermi particles, in a gas of quantum Bose particles and in a classical gas of particles, respectively [46]. In the NGF model, when the inverse temperature is sufficiently low, $\beta < \beta_c$, the generalized degree distributions of NGFs reach a stationary state in which the average generalized degrees of faces with given energy $\varepsilon_\alpha = \varepsilon$ follows the Fermi–Dirac $n_F(\varepsilon, \mu)$, the Boltzmann $n_Z(\varepsilon, \mu)$ or the Bose–Einstein distributions $n_B(\varepsilon, \mu)$, depending on the dimension d of the simplicial complex, on the dimension m of the faces and on the flavor s. Specifically, the NGFs with $0 < \beta < \beta_c$ have generalized degrees satisfying

$$\langle [k_{d,m}(\alpha) - 1] | \varepsilon_\alpha = \varepsilon \rangle = \begin{cases} n_F(\varepsilon, \mu_{d,m}(s)) & \text{for } d - m + s = 0 \\ n_Z(\varepsilon, \mu_{d,m}(s)) & \text{for } d - m + s = 1 \\ A_{d,m}^{[s]} n_B(\varepsilon, \mu_{d,m}(s)) & \text{for } d - m + s > 1, \end{cases} \tag{5.28}$$

where $A_{d,m}^{[s]} = (d-m)/(d-m-1+s)$. Table 3 summarizes these results in a concise way. From this table it can be clearly shown that for $d = 3$ dimensional NGFs with flavor $s = -1$ the triangles, the links and the nodes have the expected generalized degree minus one that follows the Fermi–Dirac, the

Table 3 The average $\langle k_{d,m} - 1|\varepsilon \rangle$ of the generalized degrees $k_{d,m} - 1$ of m-faces with energy ε in a d-dimensional NGF of flavor s follows either Fermi–Dirac, Boltzmann or Bose–Einstein statistics, depending on the values of the dimensions d and m.

Flavor	$s = -1$	$s = 0$	$s = 1$
$m = d - 1$	Fermi–Dirac	Boltzmann	Bose–Einstein
$m = d - 2$	Boltzmann	Bose–Einstein	Bose–Einstein
$m \leq d - 3$	Bose–Einstein	Bose–Einstein	Bose–Einstein

Source: Reprinted table with permission from [29] ©Copyright (2016) by the American Physical Society.

Boltzman and the Bose–Einstein distributions, respectively. Therefore the statistical properties of the faces of the NGF have a notable topological nature, i.e. they change significantly with the dimension m of the faces considered. This result generalizes the result obtained in the Bianconi–Barabási model [95], which coincides with the NGF of dimension $d = 1$ and $s = 1$. Indeed, in the Bianconi–Barabási network model the average degree of the node $\langle k_r \rangle = \langle k_{1,0}([r]) \rangle$ with energy $\varepsilon_{[r]} = \varepsilon$ follows the Bose–Einstein distribution and therefore is given by

$$\langle [k_{1,0}([r]) - 1]|\varepsilon_{[r]} = \varepsilon \rangle = n_B(\varepsilon, \mu_{1,0}). \tag{5.29}$$

The chemical potentials $\mu_{d,m}(s)$ in Eq. (5.28) and Eq. (5.29) are self-consistent parameters that must satisfy the condition

$$\left\langle \frac{\sum_{\alpha \in S_{d,m}(\mathcal{K})} k_{d,m}(\alpha)}{N_{[m]}} \right\rangle = \frac{d+1}{m+1}. \tag{5.30}$$

The value of the inverse temperature at which these conditions can no longer be satisfied indicates the critical temperature β_c. At β_c there is a topological phase transition characterized by the lack of a stationary state for the generalized degree distributions, which we will describe in the next paragraph.

The NGF model can be also generalized by considering fractional negative values of the flavor $s = -1/\bar{k}$ which enforce an upper bound of $\bar{k} + 1$ to the generalized degree of the $(d - 1)$-dimensional faces [38]. The statistical properties of this variant of the NGF are summarized in Table 4. Interestingly, in this case the $(d - 1)$-faces of the simplicial complex have $\langle k_{d,m} - 1|\varepsilon \rangle$ that follows the Fermi–Dirac statistics, while any face of dimension $m \leq d - 2$ has $\langle k_{d,m} - 1|\varepsilon \rangle$ that follows the Bose–Einstein distribution. In other words there are no faces that follow the (classical) Boltzman distribution.

Table 4 Distribution of generalized degrees of faces of dimension m in a d-dimensional NGF of flavor $s = -1/\bar{k}$ at $\beta = 0$. Only for $d - 2m \geq 2 + 3/\bar{k}$ are the power-law distributions scale-free, i.e. the second moment of the distribution diverges. The average $\langle k_{d,m} - 1|\varepsilon \rangle$ of the generalized degrees $k_{d,m}$ of m-faces with energy ε minus one in a d-dimensional NGF of flavor $s = -1/\bar{k}$ and inverse temperature $\beta > 0$ follows either Fermi–Dirac or Bose–Einstein statistics, depending on the values of the dimensions d and m.

Flavor	Generalized degree distribution	Statistics
$m = d - 1$	Bounded $k \leq \bar{k} + 1$	Fermi–Dirac
$m \leq d - 2$	Power-law	Bose–Einstein

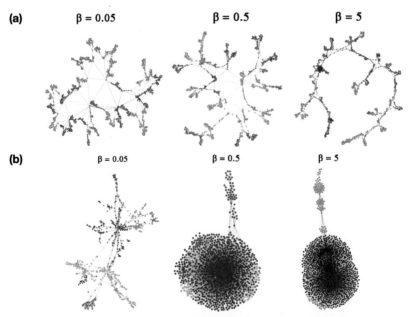

Figure 38 A visualization of the NGF with fitness of the faces as a function of the inverse temperature β with the same number of nodes N. Panel (a) shows the NGF simplicial complexes of dimension $d = 2$ and flavor $s = -1$. In this case the diameter of the network skeleton grows with the inverse temperature β. Panel (b) shows NGF simplicial complexes of dimension $d = 2$ and flavor $s = 1$. In this case the diameter of the network skeleton decreases with the inverse temperature β.

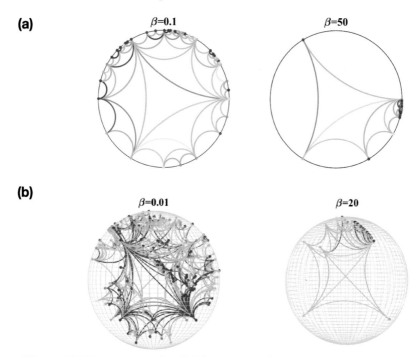

Figure 39 The emergent simplicial geometry of NGFs with fitness of the faces and flavor $s = -1$ is shown across the topological phase transition for dimension $d = 2$ (panel a) and dimension $d = 3$ (panel b). While in the high-temperature regime the network skeleton evolves homogeneously in the hyperbolic space, in the low temperature regime it evolves in one direction mostly, forming a so-called spine.

Source: Reprinted from Ref. [30].

5.5.4 Topological Phase Transition for $\beta > \beta_c$

For $\beta > \beta_c$ the NGFs display a phase transition characterized by the fact that the generalized degree distributions do not have a stationary state. This phase, called the *low-temperature phase*, corresponds to the Bose–Einstein condensate phase for the Bianconi–Barabási model (the NGF of dimension $d = 1$ and flavor $s = 1$). In the Bose–Einstein-condensate phase network, the NGF with $d = 1$ and $s = 1$ is characterized by the emergence of super-hubs having a degree growing linearly with the network size, with at most some logarithmic corrections. Additionally the Bose–Einstein condensate phase is characterized by the extremely slow growth of the diameter with the network size, i.e. a growth that is much slower than logarithmic. For general NGFs the low temperature phases are still not completely investigated. The numerical results [109] show an important difference between the low-temperature phase of NGFs with different flavors. For flavor $s = 0$ and $s = 1$ the diameter grows much slower than logarithmically with the network size, while for $s = -1$ the diameter grows as a finite power of the network size [109] (see for instance Figure 38 for a visualization of the $d = 2$ NGF with $s = -1$ and $s = 1$). In Figure 39 we show the emergent geometry of the NGF model with flavor $s = -1$ and dimension $d = 2$ and $d = 3$ below and above the topological phase transition. It is apparent that in the low-temperature phase, the evolution of the simplicial complex displays a broken symmetry and the growth occurs mostly in one direction, generating a so-called spine. Interestingly, in the low-temperature regime, the spectral properties of the NGF also change and the scaling of the density of the eigenvalues of the Laplacians deviates from power-law, in such a way that the Laplacians are no longer characterized by a finite (scale-independent) spectral dimension.

6 Higher-Order Dynamics: Synchronization

6.1 Simplicial Synchronization

This section is the first of three devoted to higher-order dynamics and discusses the vibrant field of higher-order synchronization. Synchronization is a fundamental dynamical process that describes how oscillators that are in isolation have different intrinsic frequencies, and start to display a coherent motion when their interaction is strong enough, i.e. when their coupling is above the so-called synchronization threshold. This important collective phenomenon is found across many different natural systems [110] and describes very different phenomena, including brain dynamics, the simultaneous flashing of fireflies in tropical forests, as well as the dynamical behavior of Josephson junctions.

A pivotal reference model to study synchronization is the Kuramoto model proposed by Yoshiki Kuramoto in 1975 [111]. This model is fully understood where the network of interactions between the oscillators is fully connected. In the past 20 years, the investigation of synchronization and the Kuramoto model on complex networks has been a very active research area with many interesting results [112–115]. Despite the model not yet being amenable to a complete analytical treatment in a generic complex network, important effects due to the small-world nature of the vast majority of real networks and the ubiquitous presence of a broad (scale-free) degree distribution have been pointed out [114, 116].

Synchronization is currently one of the most investigated dynamical processes in simplicial complexes [117, 118]. This very active research field has revealed how simplicial synchronization is strongly affected by the simplicial network geometry and topology.

In the Kuramoto model, each node is associated with a phase (an oscillator) and the network defines the set of pairwise interactions between neighboring oscillators. Interestingly, if the network of interactions is the skeleton of a simplicial network geometry and displays a finite spectral dimension, the dynamics can be dominated by dynamical fluctuations [35, 36]. Therefore if the spectral dimension is not large enough, the synchronized state might never be reached in the limit of large network size. This is the case, for instance, for a Network Geometry with Flavor with spectral dimension lower or equal to four. In this case it is possible to observe very significant spatio-temporal fluctuations of the order parameter. This important effect of the spectral dimension on the synchronization properties of a network skeleton of a simplicial complex can be related to recent experiments [119] showing that neuronal cultures grown on substrates of different dimension ($d = 2$ glass slices or $d = 3$ scaffolds formed by nanotube sponges) have different synchronization properties.

On a simplicial complex, the Kuramoto model can be also modified in order to take into account many-body interactions between the nodes of each simplex. In this approach, the pairwise interactions of the standard Kuramoto model defined on a network are substituted by $(d + 1)$-body interactions for the nodes of a pure d-dimensional simplicial complex. Despite the linearized dynamics being described by a weighted graph Laplacian including exclusively pairwise interactions, the full non-linear dynamics of this variation of the Kuramoto model displays a discontinuous de-synchronization transition, which is a distinct property deriving from the many-body interactions.

Finally, and importantly, simplicial topology can be used to propose topological synchronization that treats synchronization of topological signals. Indeed, simplicial topology allows mathematicians and network scientists to treat dynamical signals that are defined not only on the nodes of the simplicial complex but also on its higher-dimensional simplices [27, 28]. For instance, topological signals defined on links can indicate fluxes.

The higher-order Kuramoto model and the explosive higher-order Kuramoto model [27] display continuous and discontinuous synchronization transitions of topological signals, respectively, and and have the potential to transform the investigation of dynamical synchronization processes in neuroscience. Although there is not yet experimental evidence for synchronization of topological signals, this mathematical framework reveals that this synchronization is a phenomenon that is only noticeable if the correct topological filters are applied to the data. Therefore it is possible that if the experimental data are processed using the opportune topological transformations, the higher-order topological synchronization might be observed in brain data and in data from biological transport systems.

In this section we will summarize all these results on simplicial synchronization, emphasizing the central role that simplicial network topology and geometry play in determining simplicial dynamics.

Given the space limitation, we will not be able to treat here all the different approaches proposed to characterize the synchronization transitions in higher-order networks, such as those based on the master stability function [120, 121] and cluster synchronization [122, 123].

6.2 Kuramoto Model on Simple Networks

The Kuramoto model [111, 112, 114, 115] describes the synchronization transitions occurring among the phases associated with the nodes of a network. In the absence of interactions, the phases oscillate independently with internal frequencies that are distributed randomly. Therefore, in this case every oscillator has independent dynamics characterized by a different period. It follows that in the absence of interactions there is no collective motion of the phases and therefore no synchronization. When the interactions between the phases, modeled by the network links, are turned on, the phases evolve in time taking into account their own intrinsic frequency but also the coupling with their neighbor nodes, whose intensity is modulated by the coupling constant σ. If the coupling constant is small, the dynamics of the phase remain incoherent, but if the coupling constant is above the synchronization threshold, a phase transition occurs and most of the phases start to oscillate at the same frequency in a collective motion (see Figure 40).

Given a network of N nodes defined by an adjacency matrix $\mathbf{A}$, the Kuramoto model assigns to each node r of the network a phase ϕ_r evolving in time as

$$\dot{\phi}_r = \omega_r + \sigma \sum_{s=1}^{N} A_{rs} \sin(\phi_s - \phi_r), \qquad (6.1)$$

where σ is the coupling constant and ω_r are the internal frequencies of the nodes drawn randomly from a distribution $g(\omega)$ with average Ω_0. The synchronization phase transition is captured by the global order parameter R defined as

$$R = \frac{1}{N} \left| \sum_{r=1}^{N} e^{i\phi_r} \right|. \qquad (6.2)$$

The role of R is to describe the synchronization transition defined in the limit $N \to \infty$ and characterized by the *synchronization threshold* σ_c such that for $|\sigma - \sigma_c| \ll 1$ we observe

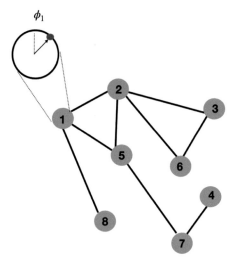

Figure 40 Schematic representation of the Kuramoto model on a network. Each node of the network is assigned a dynamical variable (a phase) that oscillates at its own frequency in the absence of interactions. The network represents the interactions between neighboring nodes.

Source: Reprinted from [120].

$$R = \begin{cases} 0 & \text{for} \quad \sigma \leq \sigma_c, \\ C(\sigma - \sigma_c)^\beta & \text{for} \quad \sigma > \sigma_c. \end{cases} \qquad (6.3)$$

On a fully connected network, the synchronization transition is fully understood theoretically. On fully connected networks the coupling constant must be rescaled by the number of nodes, i.e. $\sigma \to \sigma/N$. With this rescaling of the coupling constant, the synchronization transition on a fully connected network is known to be continuous with the synchronization threshold given by $\sigma_c = 2/[\pi g(\Omega_0)]$ (see Figure 41).

6.3 Kuramoto Models on Simplicial Network Geometry

Simplicial network geometry characterizes simplicial complexes and their network skeleton according to their spectral dimension. Indeed, while random networks are typically characterized by a finite spectral gap, in simplicial network geometry the spectral gap closes and the density of eigenvalues displays the characteristic power-law scaling defined in Eq. (4.11). In this section we will discuss how the presence of a spectral dimension strongly affects the synchronization properties of the network skeleton of a simplicial complex as revealed in Ref. [36]. The relation between the dynamics of the Kuramoto

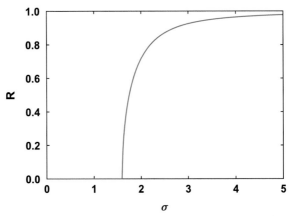

Figure 41 The Kuramoto model on an infinite fully connected network is characterized by a continuous phase transition in which R, the synchronization order parameter, goes from a zero to a non-zero value. Here the integral frequencies are taken from a Gaussian distribution $\omega \sim \mathcal{N}(\Omega_0, 1)$.

model and the geometry of the network skeleton of a simplicial complex is fully encoded in the spectral properties of the graph Laplacian. In fact, if we linearize the Kuramoto model close to the synchronized state it is possible to appreciate the relation between the Kuramoto model and the graph Laplacian. Close to the synchronization transition, the phases of the nodes have small fluctuations around a dynamical state in which they all oscillate at the same frequency Ω_0, i.e.

$$\phi_r = \Omega_0 t + \psi_r, \quad \text{with } \psi_r \ll 1. \tag{6.4}$$

The Kuramoto dynamics can be then linearized, obtaining the equation

$$\dot{\psi}_r = \omega_r - \Omega_0 - \sigma \sum_{s=1}^{N} [L_{[0]}]_{rs} \psi_s, \tag{6.5}$$

which clearly demonstrates the role that the graph Laplacian plays in determining the stability of the synchronized phase. By studying these linearized dynamics it is possible to conclude [36] that the spectral dimension of the graph Laplacian has an important role in guaranteeing the stability of the synchronized phase.

SYNCHRONIZATION AND SPECTRAL DIMENSION [36]

The Kuramoto model cannot synchronize networks with finite spectral dimension

$$d_S \leq 2 \tag{6.6}$$

and in these networks it is found in the incoherent state for every value of the coupling constant σ. Networks with finite spectral dimension

$$2 < d_S \leq 4 \tag{6.7}$$

can only display a incoherent state for entrained phases. The synchronized phase can be only achieved in the large network limit as long as the networks have spectral dimension

$$d_S > 4. \tag{6.8}$$

Note that on very heterogeneous networks, if we desire to screen out the effect of the heterogeneity of the degrees, a very common modification of Kuramoto dynamics is to consider the system of equations

$$\dot{\phi}_r = \omega_r + \sigma \sum_{s=1}^{N} \frac{A_{rs}}{k_r} \sin{(\phi_s - \phi_r)}, \tag{6.9}$$

where k_r is the degree of node r. Also interesting for these dynamics is that the synchronized phase is not achieved in the large network limit if the spectral dimension $d_S \leq 4$. However, if we consider the normalized Kuramoto dynamics the spectral dimension d_S is the spectral dimension of the normalized Laplacian $\tilde{\mathbf{L}}$ or elements $\tilde{L}_{rs} = \delta(r,s) - A_{rs}/k_r$. In Ref. [35, 36] the case of NGFs with flavor $s = -1$ and dimension $d \in \{2, 3, 4\}$ has been considered. These NGFs have a network skeleton with spectral dimension approximately equal to $d_S \sim d$, therefore they cannot sustain a synchronized phase in the limit $N \to \infty$. For NGFs of finite network size N, it is possible to observe a region of strong spatio-temporal fluctuations of the order parameter (see Figure 42). The size of this region increases as the network size N increases.

These results can shed new light on a recent finding in the field of neuroscience demonstrating the important role that network geometry has in the synchronization properties of neuronal cultures. In Ref. [119] it has been shown that neuronal cultures growing in 2-dimensional slices sustain dynamical states in which neurons are significantly less correlated than neurons growing in 3-dimensional scaffolds. This result implies that the network geometry of the neuronal network has important effects on the synchronization properties of the neuronal culture and that increasing the dimensionality of the neuronal network increases the coherence of the dynamics in line with the findings discussed in this section.

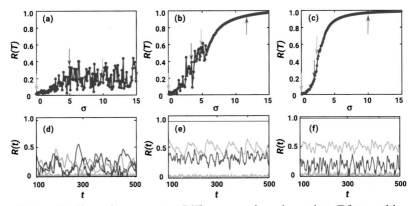

Figure 42 The order parameter $R(T)$ measured at a large time T for tunable values of the coupling constant σ is shown for the NGF (neutral model) with flavor $s = -1$ and dimension $d = 2$ (panel a) $d = 3$ (panel c) and dimension $d = 4$ (panel c). Panels (d)-(e)-(f) represent a typical time series $R(t)$ corresponding to the values of σ indicated with an arrow in the panels (a)-(b)-(c), respectively. These results highlight the important temporal fluctuations of the order parameter induced by the spectral dimension $d_S \simeq d$ with $d_S \leq 4$.

Source: Reprinted from [35].

6.4 Kuramoto Model with Many-Body Interactions

Simplicial complexes are known to describe many-body interactions. Therefore a central question is how to modify the Kuramoto model to include many-body interactions? Interesting work proposes to consider a Kuramoto model in which phases are always assigned to the nodes of the simplicial complex, but the interactions can include more than two nodes. There is no unique choice for these many-body interactions, however Ref. [13] proposes considering the following set of equations for a fully connected 2-dimensional simplicial complex,

$$\dot{\phi}_r = \omega_r + \frac{K}{N^2} \sum_{s=1}^{N} \sum_{q=1}^{N} \sin(\phi_s + \phi_q - 2\phi_r). \tag{6.10}$$

Here ω_r is the internal frequency of the nodes drawn from a distribution $g(\omega)$ and the normalized coupling constant is indicated by K. This higher-order synchronization model admits order parameters r_1 and r_2 defined as

$$r_z = \frac{1}{N} \sum_{r=1}^{N} e^{iz\phi_r}, \tag{6.11}$$

for $z \in \{1, 2\}$. The synchronization properties of this model can be studied by investigating the forward and backward transitions (see Figure 43), which are

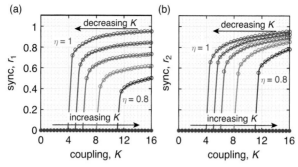

Figure 43 The order parameters r_1 and r_2 versus the normalized coupling constant K along the backward (decreasing K) and forward (increasing K) transition for the Kuramoto dynamics defined by Eq. (6.10). Here the different curves indicate initial conditions in which the phases are aligned with probability η.

Source: Reprinted figure with permission from [13] ©(2019) by the American Physical Society.

amenable to analytic investigation due to the fact that the underlying simplicial complex is fully connected. In fact, while from random initial conditions the model does not display a synchronization transition, if one starts from aligned or partially aligned phases and dynamically reduces the coupling constant K, a discontinuous de-synchronization transition is observed (this transition is also called the backward transition). If instead one starts from a incoherent state at a very small value of the coupling constant K, it is not possible to observe a transition as the coupling constant K is increased (we therefore say that there is no forward transition). This model has been generalized in different directions [125, 126], including a generalization to oscillators defined by the rotation of a unitary vector in dimension larger than 2 [127]. Among other variations, here we cite the relevant extension to simplicial complexes of arbitrary dimension d [128]. Assuming that all the m-dimensional simplices of the simplicial complex are captured by the tensor $\mathbf{a}^{[m]}$ Ref. [128] proposed to investigate the model

$$\dot{\phi}_r = \omega + \sum_{m=2}^{d} \frac{K}{N^m} \sum_{\alpha=[r,v_1,v_2,\ldots,v_m]\in S_m(\mathcal{K})} a_{\alpha}^{[m]} \sin\left(\sum_{j=1}^{m} \phi_{v_j} - m\phi_r\right), \quad (6.12)$$

with all the nodes having identical frequencies ω, which allows analytical predictions to be performed on the model. Interestingly, both the original model and this latter generalization have a linearized dynamics that is dictated by a weighted Laplacian matrix $\hat{\mathbf{L}}$ [128]. For the latter general model, the weighted

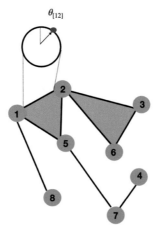

Figure 44 A schematic representation of the topological synchronization captured by the higher-order Kuramoto model associating a phase to each m dimensional simplex. Here $m = 1$.

Source: Reprinted from [124].

Laplacian matrix $\hat{L}$ has elements $\hat{L}_{rs} = W_r \delta(r, s) - A_{rs} w_{rs}$, where A_{rs} is the element of the adjacency matrix of the network skeleton of the simplicial complex, w_{rs} is the weight of the link $[r, s]$ given by the sum of the generalized degrees of the link $[r, s]$ i.e. $w_{rs} = \sum_{m=1}^{d} k_{m,1}([r, s])$ and $W_r = \sum_{s=1}^{N} w_{rs}$ is the strength of node r. This implies that the linearized model of these variants of the Kuramoto model is a pairwise diffusion model, so all the non-trivial effects related to the many-body interactions are non-linear.

6.5 Higher-Order Kuramoto Model

Simplicial complex topology allows for the formulation of a higher-order Kuramoto model [27] that describes the synchronization of topological signals (see Figure 44). In the higher-order Kuramoto model the dynamical variables are phases that can be associated with any m-dimensional faces of the simplicial complex with $m \geq 1$, such as links ($m = 1$) or triangles ($m = 2$). The higher-order Kuramoto model uses topology (see Section 3.2 for an introduction) to naturally define higher-order interactions among the topological signals. In this way the many-body interactions between the topological signals are uniquely defined. Interestingly, the linearized version of this model still retains its higher-order character as it is determined by the higher-order Laplacian. The higher-order Kuramoto model displays a continuous topological synchronization transition; however, the model can be modified to obtain the explosive higher-order Kuramoto model that displays a discontinuous phase transition.

In order to introduce the equations dictating the dynamical evolution of the higher-order Kuramoto model let us first revisit the equations of the standard Kuramoto model on a network. The standard Kuramoto model can be expressed in terms of the incidence matrix $\mathbf{B}_{[1]}$ (defined in Section 3.2) as

$$\dot{\boldsymbol{\phi}} = \boldsymbol{\omega} - \sigma \mathbf{B}_{[1]} \sin \mathbf{B}_{[1]}^{\top} \boldsymbol{\phi}, \tag{6.13}$$

where we have used the notation $\sin \mathbf{x}$ to indicate the column vector, with the sine function taken element-wise and where

$$\boldsymbol{\phi} = (\phi_1, \phi_2, \dots, \phi_r \dots)^{\top}, \tag{6.14}$$
$$\boldsymbol{\omega} = (\omega_1, \omega_2, \dots, \omega_r \dots)^{\top} \tag{6.15}$$

indicate the phases of the nodes and their intrinsic frequencies, respectively. For the higher-order Kuramoto model we associated with each m-dimensional simplex α a phase θ_α. For instance, for $m = 1$ we might associate an oscillating flux with each link (see Figure 44). The column vector of phases of the m-dimensional simplices is indicated by $\boldsymbol{\theta}$ and its generic element is θ_α. The higher-order Kuramoto model is defined as

$$\dot{\boldsymbol{\theta}} = \hat{\boldsymbol{\omega}} - \sigma \mathbf{B}_{[n+1]} \sin \mathbf{B}_{[n+1]}^{\top} \boldsymbol{\theta} - \sigma \mathbf{B}_{[n]}^{\top} \sin \mathbf{B}_{[n]} \boldsymbol{\theta}, \tag{6.16}$$

where the internal frequencies of the phases are indicated by the column vector $\hat{\boldsymbol{\omega}}$ having a generic element given by $\hat{\omega}_\alpha$ drawn from a random distribution $g(\hat{\omega})$ (for instance, we can take the Gaussian distribution $\hat{\omega}_\alpha \sim \mathcal{N}(\Omega_0, 1)$). The higher-order topological Kuramoto model naturally generalizes the standard Kuramoto model, at the same time prescribing a way to define higher-order interactions that are determined by the simplicial network topology. In particular the linearized version of the higher-order Kuramoto model is given by the m-order Laplacian matrix, i.e.

$$\dot{\boldsymbol{\theta}} = \hat{\boldsymbol{\omega}} - \sigma \mathbf{L}_{[m]} \boldsymbol{\theta}. \tag{6.17}$$

Therefore this higher-order Kuramoto model also retains its higher-order character when it is linearized and uniquely determines the choice of the form of the many-body interactions, as this is the natural choice dictated by simplicial topology.

Having defined the higher-order Kuramoto model on m-simplices, a key question is: What are the dynamics induced on $(m-1)$ faces and $(m+1)$ faces? For instance, if we study the synchronization of oscillating fluxes defined on the links ($m = 1$), what are the dynamics induced on nodes ($m = 0$) and triangles ($m = 2$)? A natural way to project the dynamics is to use the incidence matrices and define the projected dynamics $\theta^{[-]}$ and $\theta^{[+]}$ on $(m-1)$ and $(m+1)$ faces given by

$$\theta^{[-]} = \mathbf{B}_{[n]}\theta, \tag{6.18}$$

$$\theta^{[+]} = \mathbf{B}_{[n+1]}^{\top}\theta. \tag{6.19}$$

In the case $m = 1$, $\theta^{[-]}$ defined on the nodes of the simplicial complex indicates the discrete divergence of the original signal defined on the links. Moreover, $\theta^{[+]}$ defined on the triangles of the simplicial complex indicates the discrete curl of the original signal defined on the links. Thanks to the Hodge decomposition (see Section 3.2), the projected dynamics on the $(m - 1)$ and $(m + 1)$ faces decouple, as we have

$$\dot{\theta}^{[+]} = \mathbf{B}_{[n+1]}^{\top}\hat{\omega} - \sigma\, \mathbf{L}_{[n+1]}^{[down]} \sin(\theta^{[+]}), \tag{6.20}$$

$$\dot{\theta}^{[-]} = \mathbf{B}_{[n]}\hat{\omega} - \sigma\mathbf{L}_{[n-1]}^{[up]} \sin(\theta^{[-]}). \tag{6.21}$$

It is therefore natural to define two distinct order parameters, one for each of the projected dynamics,

$$R^{[+]} = \frac{1}{N_{[n+1]}} \left| \sum_{\alpha \in S_{d+1}(\mathcal{K})} e^{i\theta_{\alpha}^{[+]}} \right|$$

$$R^{[-]} = \frac{1}{N_{[n-1]}} \left| \sum_{\alpha \in S_{d-1}(\mathcal{K})} e^{i\theta_{\alpha}^{[-]}} \right|. \tag{6.22}$$

These order parameters undergo a synchronization transition at $\sigma_c = 0$ (see Figure 45), as confirmed by a phenomenological model of this transition [27] (see Figure 47). However, the naïve order parameter of the topological signals R given by

$$R = \frac{1}{N_{[m]}} \left| \sum_{\alpha \in S_d(\mathcal{K})} e^{i\theta_{\alpha}} \right|, \tag{6.23}$$

does not reveal any sign of the topological synchronization and always remains close to zero for every value of the coupling constant σ. This means that only if we perform the correct topological transformation of the topological signal can we reveal higher-order synchronization.

Higher-order topological synchronization can undergo a discontinuous transition when the *explosive higher-order Kuramoto model* is considered [27]. The explosive higher-order Kuramoto model obeys the system of equations

$$\dot{\theta} = \omega - \sigma R^{[-]}\mathbf{B}_{[n+1]} \sin\, \mathbf{B}_{[n+1]}^{\top}\theta - \sigma R^{[+]}\mathbf{B}_{[n]}^{\top} \sin\, \mathbf{B}_{[n]}\theta, \tag{6.24}$$

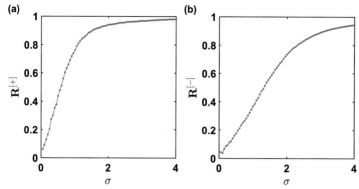

Figure 45 The higher-order Kuramoto model describes the synchronization of topological signals defined on m-dimensional simplices. The order parameter of the dynamics are $R^{[+]}$ and $R^{[-]}$ and reveal the synchronization of the projected dynamics on $(m+1)$- and $(m-1)$-dimensional simplices, respectively. In the absence of adaptive coupling of the order parameter, the topological synchronization is continuous and occurs at $\sigma_c = 0$. Here the result is obtained in a configuration model of simplicial complexes and m is taken to be $m = 1$.

Source: Adapted from [27].

therefore the coupling constant σ is adaptively multiplied by the order parameters $R^{[-]}$, $R^{[+]}$. The explosive higher-order Kuramoto model allows coupled dynamics of the projection of the phases on the $(m+1)$ and $(m-1)$ faces given by

$$\dot{\theta}^{[+]} = \mathbf{B}_{[n+1]}^{\mathsf{T}}\omega - \sigma R^{[-]}\,\mathbf{L}_{[n+1]}^{[down]}\sin(\theta^{[+]}), \tag{6.25}$$

$$\dot{\theta}^{[-]} = \mathbf{B}_{[n]}\omega - \sigma R^{[+]}\,\mathbf{L}_{[n-1]}^{[up]}\sin(\theta^{[-]}), \tag{6.26}$$

where the coupling is due to the adaptive coupling modulating the coupling constant σ with the global order parameters $R^{[-]}$ and $R^{[+]}$. This adaptive coupling, reminiscent of the coupling considered in Ref. [129] for multiplex networks, causes a discontinuous phase transition at a non-zero value of the coupling constant σ_c (see Figures 46 and 47). Interestingly, the discussed phenomenology of the higher-order topological synchronization, confirmed by the phenomenological theory of the process (see Figure 47), is observed on very different simplicial complexes, including the configuration model of simplicial complex NGFs and clique complexes or real connectomes [27].

In conclusion, higher-order topological synchronization combines Hodge theory with the theory of dynamical systems to shed light on higher-order synchronization. With this theoretical framework it is possible to treat synchronization of topological dynamical signals associated with links, such as

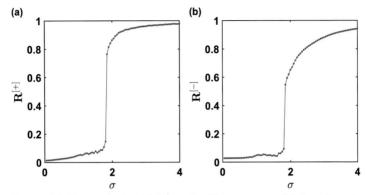

Figure 46 The explosive higher-order Kuramoto model describes the synchronization of topological signals defined on m-dimensional simplices. The order parameters of the dynamics $R^{[+]}$ and $R^{[-]}$ reveal the synchronization of the projected dynamics on $(m + 1)$ and $(m - 1)$-dimensional simplices, which is discontinuous and occurs at a non-zero synchronization threshold σ_c. Here the results are obtained in a configuration model of simplicial complexes and m is taken to be $m = 1$.

Source: Adapted from [27].

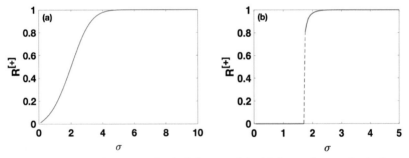

Figure 47 A phenomenological theory of the higher-order topological Kuramoto model predicts, in the absence of adaptive coupling, a continuous synchronization transition at $\sigma_c = 0$ (panel a) and, in the presence of adaptive coupling, a discontinuous phase transition at a non-zero value of the coupling constant σ_c (panel b).

Source: Adapted from [27].

fluxes, or with triangles or other higher-order simplices. The simple higher-order Kuramoto model of m-dimensional topological signals induces dynamics on $m + 1$ and $m - 1$ faces that is uncoupled and synchronizes at a continuous synchronization transition with $\sigma_c = 0$. The explosive higher-order Kuramoto model couples the projected dynamics on $m + 1$ and $m - 1$ faces, inducing a discontinuous transition. Finally, we observe that topological synchronization reveals that topological signals can undergo a synchronization transition, but this synchronization can be unnoticed if the correct topological transformations are not performed. Therefore, topology suggests a filtering of the topological signals that is the equivalent of a Fourier transform and can reveal the topological synchronization in real systems such as biological transport networks and the brain. Interestingly, in Ref. [130] this model has been adapted to propose a consensus model.

6.6 Coupling Topological Signals of Different Dimensions

We conclude this section by mentioning that topological signals of different dimensions can also co-exist and interact with each other in the same simplicial complex. As a first example of topological synchronization dynamics that couple dynamical signals, we mention here the coupled dynamics of topological signals of nodes and links [124] dictated by the system of differential equations

$$\dot{\phi} = \omega - \sigma R_1^{[-]} \mathbf{B}_{[1]} \sin \, \mathbf{B}_{[1]}^\top \phi$$
$$\dot{\theta} = \hat{\omega} - \sigma R_1^{[-]} \mathbf{B}_{[n+1]} \sin \, \mathbf{B}_{[n+1]}^\top \theta - \sigma R_0 R_1^{[+]} \mathbf{B}_{[n]}^\top \sin \, \mathbf{B}_{[n]} \theta. \qquad (6.27)$$

Here ϕ indicates the vector of phases associated with the nodes of the simplicial complex, and θ indicates the vector of phases associated with the links of the simplicial complex. Similarly ω indicates the vector of the internal frequencies associated with the nodes, while $\hat{\omega}$ indicates the vector of internal frequencies associated with the links of the simplicial complex. The adaptive coupling of these dynamics is modulated by the order parameters $R^{[+]}$ and $R^{[-]}$ of the phases of the links projected on nodes and triangles, respectively, and the standard order parameter R_0 of the Kuramoto dynamics defined on the nodes of the simplicial complex. This topological synchronization of coupled topological signals displays a discontinuous transition indicated by a discontinuity of all three order parameters $R^{[+]}$, $R^{[-]}$ and R_0 occurring for the same value of the coupling constant σ_c.

7 Higher-Order Dynamics: Percolation

7.1 Interplay Between Percolation and Simplicial Network Geometry and Topology

Simplicial network geometry and topology can have a very significant effect on percolation. This effect is particularly relevant when the simplicial complexes are hyperbolic. In fact, while in non-hyperbolic networks percolation is known to display only one phase transition, in hyperbolic networks at least two transitions are observed, corresponding to the lower and the upper percolation thresholds [131]. In this section we will cover link percolation in hyperbolic simplicial and cell complexes, including manifolds and more general structures. Interestingly, in these structures the upper percolation transition can be discontinuous [132, 133], or continuous with non-trivial critical behavior, depending on the underlying simplicial geometry [133–135]. This important result shows the extremely significant implications of considering a dynamical process such as percolation on the network skeleton of hyperbolic simplicial complexes, revealing the important consequences of network geometry on dynamics. Finally, in this section we will introduce topological percolation [37] which studies m-connectedness in the presence of topological damage. In a network there are only two possible models of percolation, node and link, that characterize the 0-connectedness of the networks when nodes or links are randomly damaged. On the other hand, in simplicial complexes topological damage can also target higher-order simplices like triangles and tetrahedra. Topological percolation therefore refers to models of percolation studying m-connectedness with m in the range $0 \leq m < d$ when topological damage is targeting either m or $m + 1$ simplices. By considering the analogy of a geographical map, the damage to a link can be represented as a river that inhibits diffusion from one region of land (triangle) to another. Interestingly, we will show that the critical properties of different topological percolation models defined on the same simplicial complex can be significantly different. Moreover, we will provide evidence that the properties of higher-order topological percolation cannot be predicted starting from the properties of node or link percolation on the same simplicial complex.

A few works [136, 137] have also treated *homological percolation*, which describes the emergence of giant cycles in simplicial complexes. Another line of research in which there is growing interest is at the interface between combinatorial complexity, statistical mechanics and computer science, and concerns hypergraph covering problems [138, 139]. We regret to be unable to treat these interesting processes here due to space limitations.

7.2 Link Percolation in Hyperbolic Manifolds And Pseudo-Fractal Networks

In order to investigate the interplay between percolation and simplicial network geometry, here we focus on link percolation on the network skeleton of $d = 2$ dimensional simplicial complexes with hyperbolic geometry [132–135]. Link percolation predicts the size of the giant component of a network when links are removed randomly with probability $1 - p$. Link percolation in random networks and in general in networks with a finite spectral gap has a single percolation threshold for $p = p_c$.

- For $p < p_c$ the 0-connected components are all finite in the limit $N \to \infty$.
- For $p = p_c$ the giant component is infinite in the limit $N \to \infty$; however, it includes an infinitesimal fraction of all the nodes of the network, i.e. it is subextensive.
- For $p > p_c$ an extensive giant component emerges, including a finite fraction of all the nodes N of the network, i.e. the giant component is both infinite and extensive in the limit $N \to \infty$.

The scenario changes significantly for hyperbolic networks. In this case link percolation is known to have at least two percolation thresholds: the lower percolation threshold $p^\star$ and the upper percolation threshold p_c.

- For $p < p^\star$ all clusters are finite.
- For $p^\star \leq p \leq p_c$ the maximum cluster is infinite (i.e. it has a size that diverges as the network size N diverges) but is subextensive.
- For $p > p_c$ the maximum cluster is extensive.

Therefore, in hyperbolic networks the phase with $p^\star < p \leq p_c$ can be considered as a stretched *critical region* observed for all values of p belonging to the mentioned interval. Interestingly, while link percolation is always characterized by a second-order phase transition in networks that are not hyperbolic, in hyperbolic networks, link percolation can be discontinuous at the upper percolation threshold. This result was first demonstrated by Boettcher, Singh and Ziff in Ref. [132]. In this paper the authors fully characterized the upper percolation transition for link percolation on the network skeleton of a $d = 2$ hyperbolic simplicial complex manifold (see Figure 48 a). This manifold is constructed starting at time $t = 0$ from a single link and at each time $t > 0$ a triangle is glued to every link added at the previous time step. Therefore this is a deterministic hyperbolic manifold that constitutes a variation of the Apollonian simplicial complex treated in Appendix G. This construction of a $d = 2$ dimensional hyperbolic manifold can be generalized by considering cell complexes representing $d = 2$ dimensional manifolds [133]. This generalization consists in considering hyperbolic manifolds which describe the network skeleton of cell complexes constructed according to the following algorithm. Starting at time $t = 0$ from a single link, at each time $t > 0$ an m-polygon whose size is drawn from a distribution q_m, is attached to every link added at the previous timestep (see Figure 48 panels b and c). As a first example of link percolation on such a hyperbolic network structure, we will consider the case in which every added polygon has the same number of links m, i.e. the case in

(a)   **(b)**   **(c)**

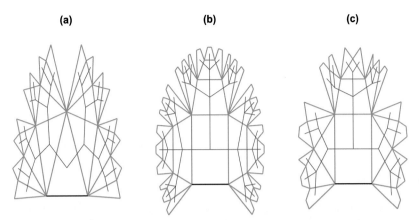

Figure 48 Schematic representation of discrete 2-dimensional hyperbolic manifolds defined by simplicial complexes (panel a) or cell complexes (panels b and c). While the cell complex of panel (b) is formed by identical m-polygons (squares) and has $q_m = \delta(m, 4)$, the cell complex in panel (c) combines polygons of size 3 and size 4, with $q_m = 1/2\delta(m, 3) + 1/2\delta(m, 4)$.

which $q_{m'} = \delta(m', m)$. In this case the link percolation at the upper percolation threshold is discontinuous, with a discontinuity in the fraction of nodes in the extensive component $P_\infty(p_c)$ that depends on m as [133]

$$P_\infty(p_c) \simeq Ce^{-\frac{m}{3(m-1)}}, \tag{7.1}$$

where C is a constant. The excellent agreement of this theoretical prediction with exact numerical results is shown in Figure 49.

In the case in which the size of each new m-polygon is drawn from a power-law distribution, $q_m = Cm^{-\gamma}$, for $m > 3$ the nature of the upper percolation transition changes. Using the renormalization group, it can be shown [133] that while for $p < p_c$ the fraction $P_\infty(p)$ of nodes in the extensive component is zero, i.e. $P_\infty(p) = 0$, for $0 < \Delta p = p - p_c \ll 1$, $P_\infty(p)$ has a scaling that depends on the value of γ and is given by

$$P_\infty(p) \simeq \begin{cases} P_\infty(p_c) + c_\gamma \Delta p \ln \Delta p & \text{for} \quad \gamma > 4, \\ P_\infty(p_c) + c_\gamma \Delta p [\ln \Delta p]^2 & \text{for} \quad \gamma = 4, \\ P_\infty(p_c) + c_\gamma [\Delta p]^{\gamma-2} & \text{for} \quad \gamma \in (3, 4), \\ c_\gamma e^{-C/[\Delta p]} & \text{for} \quad \gamma = 3, \\ c_\gamma [\Delta p]^\beta & \text{for} \quad \gamma \in (2, 3), \end{cases} \tag{7.2}$$

where $P_\infty(p_c)$ indicates the discontinuity at $p = p_c$, c_γ are constants and β is the dynamical critical exponent. Therefore only for $\gamma > 3$ is the transition

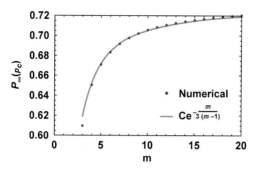

Figure 49 The discontinuity $P_\infty(p_c)$ at the upper percolation threshold p_c of link percolation on 2-dimensional hyperbolic manifolds with $q_{m'} = \delta(m, m')$ is shown here versus m. The exact numerical results (blue dots) are compared with the predicted theoretical scaling (orange line) showing excellent agreement. For simplicial complexes ($m = 3$) the discontinuity is smaller than for regular cell complexes with $m > 3$.

Source: Reprinted figure with permission from [133] ©Copyright (2019) by the American Physical Society.

discontinuous, while for $\gamma \leq 3$ the transition becomes continuous. In other words, the upper link percolation is discontinuous as long as q_m has finite first and second moments.

Using the renormalization group it is also possible [135] to characterize the universality class of link percolation for a variant of the $d = 2$ dimensional pseudo-fractal cell complexes (for pseudo-fractal simplicial complexes see Appendix G). This variant of pseudo-fractal cell complexes of dimension $d = 2$ considers cell complexes constructed iteratively starting at time $t = 0$ from a single link, and evolving iteratively in time so that at each time a new m-polygon is glued to each link of the cell complex. Link percolation in $d = 2$ dimensional simplicial complexes has an upper percolation threshold at $p_c = 0$ (implying that there is no lower percolation threshold). The nature of the phase transition is continuous but anomalous, as the fraction of nodes $P_\infty(p)$ in the extensive component scales as [135]

$$P_\infty(p) \propto p e^{-b/p^{m-2}}, \tag{7.3}$$

where $b >$ is a constant, for $p \ll 1$. This expression is interesting as it indicates a critical suppression of the giant component that is faster than any power, regardless of the value of m. We would then say that the dynamical critical exponent is infinite for any value of m. Nevertheless the exponential suppression of P_∞ depends on m. This shows clearly that the dynamical processes defined on simplicial complexes and their cell complex counterparts might be noticeably different, emphasizing the important role that network topology and geometry have in dynamical processes.

On branched hyperbolic simplicial and cell complexes that are not discrete manifolds, link percolation can display different universality classes at the upper percolation threshold. Moreover, in some cases it is possible to observe intermediate phase transitions between the lower and the upper percolation thresholds, in which the maximum size of the infinite clusters has a discontinuity but remains subextensive across the transition [134].

7.3 Topological Percolation

Having discussed the role that simplicial network geometry has in percolation, let us show the role that simplicial network topology has in determining the properties of percolation. This line of reasoning leads to the formulation of topological percolation [37].

In networks, damage can occur only on nodes or on links. In simplicial complexes topological damage can also be directed to higher-dimensional simplices, such as triangles, tetrahedra etc. This allows us to define a set of topological percolation models investigating m-connectedness in the presence of topological damage [37]. In a simplicial complex of dimension d we can define $2 \times d$ topological percolation models. Let us show how to construct these models for simplicial complexes of dimension $d = 1, 2, 3$.

- On simplicial complexes of dimension $d = 1$ we distinguish between two topological percolation problems:
  - *Link percolation*: Links are removed with probability $1 - p$. Nodes are connected to nodes through intact links.
  - *Node percolation*: Nodes are removed with probability $1 - p$. Links are connected to links through intact nodes.
- On $d = 2$ simplicial complexes we distinguish four types of topological percolation models: link percolation and node percolation defined as for $d = 1$, and additionally:
  - *Triangle percolation*: Triangles are removed with probability $1 - p$. Links are connected to links through intact triangles.
  - *Upper-link percolation:* Links are removed with probability $1 - p$. Triangles are connected to triangles though intact links.
- On $d = 3$ simplicial complexes we distinguish six types of topological percolation problems: node percolation, link percolation, triangle percolation and upper-link percolation defined as in $d = 2$ and additionally:
  - *Tetrahedron percolation*: Tetrahedra are removed with probability $1 - p$. Triangles are connected to triangles though intact tetrahedra.

(a)   (b)

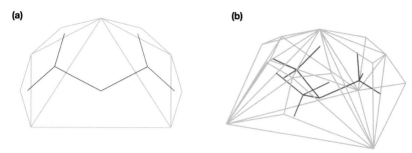

Figure 50 Schematic representation of $d = 2$ and $d = 3$ hyperbolic manifolds considered in Ref. [37] as benchmarks to study topological percolation.

Source: Reprinted figure with permission from [37] ©Copyright (2018) by the American Physical Society.

- *Upper-triangle percolation*: Triangles are removed with probability $1 - p$. Tetrahedra are connected to tetrahedra though intact triangles.

In Ref. [37] topological percolation has been formulated and studied in hyperbolic manifolds of different dimensions. This work reveals the relation between different topological percolation models studied on the same simplicial complex. Moreover, it investigates the relation between topological percolation models defined on simplicial complexes built following the same principles, but having different dimension d.

The $d = 2$ and $d = 3$ hyperbolic manifolds studied in Ref. [37] are variations of the Apollonian simplicial complexes (see Figure 50). These hyperbolic manifolds are constructed starting at time $t = 0$ from a single $(d - 1)$-dimensional simplex (i.e. a link for $d = 2$ and a triangle for $d = 3$ manifolds). Each time $t \geq 1$, a d-dimensional simplex is attached to each $(d - 1)$-dimensional face added at the previous time step. This construction generates a d-dimensional hyperbolic lattice. The lower and upper percolation thresholds for every topological percolation model defined on these hyperbolic manifolds is reported in Table 5. All upper percolation transitions occurring for $p = 1$ are trivial, as they imply that an extensive component is only achieved in the absence of topological damage. For the $d = 2$ hyperbolic manifold, the only non-trivial upper percolation transition is the discontinuous upper percolation transition of link percolation, found originally in Ref. [132] and discussed in the previous paragraph. For the $d = 3$ hyperbolic manifold, the non-trivial upper percolation transitions are the ones observed for link percolation and for triangle percolation. For link percolation, the transition is continuous at $p_c = 0$ and is characterized by an exponential suppression of the order parameter $P_\infty(p)$, [140] i.e.

Table 5 Lower $p^\star$ and upper p_c percolation thresholds for topological percolation on the $d = 2$ and $d = 3$ hyperbolic manifolds described in Figure 50.

$d = 2$	$p^\star$	p_c	Transition
Link percolation	0	$\frac{1}{2}$	Discontinuous (Non-trivial)
Triangle percolation	$\frac{1}{2}$	1	Discontinuous
Node percolation	0	1	Discontinuous
Upper link percolation	$\frac{1}{2}$	1	Discontinuous
$d = 3$	p^l	p^u	Transition
Link percolation	N/A	0	Continuous (Exponential)
Triangle percolation	0	$0.307981\ldots$	Continuous (BKT)
Tetrahedra percolation	$\frac{1}{3}$	1	Discontinuous
Node percolation	0	1	Discontinuous
Upper link percolation	0	1	Discontinuous
Upper triangle percolation	$\frac{1}{3}$	1	Discontinuous

$$P_\infty(p) \propto e^{-b/p}, \tag{7.4}$$

where $b > 0$ is a constant.

Triangle percolation is a new non-trivial topological percolation model which describes the connectivity of links though paths passing across shared triangles (see Figure 51). The upper percolation threshold of triangle percolation on $d = 3$ hyperbolic manifolds is non-trivial, as it falls into the Berezinskii–Kosterlitz–Thouless (BKT) universality class with order parameter $P_\infty(p)$ scaling as [37]

$$P_\infty(p) \propto e^{-b/|p-p_c|^{1/2}}, \tag{7.5}$$

where $b > 0$ is a constant.

From Table 5 we take the following considerations:

- For the $d = 2$ hyperbolic manifold all transitions are discontinuous, while in $d = 3$, link and triangle percolations are continuous.
- Link percolation on the $d = 2$ hyperbolic manifold displays a non-trivial discontinuous transition at the upper percolation threshold, while no such transition is observed for the $d = 3$ manifold.
- Triangle percolation for the $d = 3$ hyperbolic manifold is a BKT transition, while no such transition is observed for the $d = 2$ hyperbolic manifold.

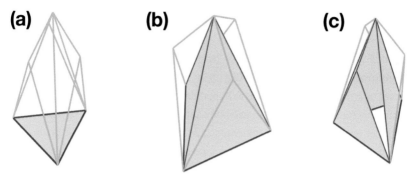

Figure 51 Schematic description of triangle percolation on the $d = 3$ hyperbolic manifold. The three links indicated in red are connected through shared undamaged (filled) triangles in three different realizations of the topological damage (panels a-b-c).

Source: Reprinted figure with permission from [37] ©Copyright (2018) by the American Physical Society.

We therefore deduce two important take-home messages:

- Node and link percolations cannot be used to predict other topological percolation problems.
- The nature of topological percolation transitions on $d = 2$ hyperbolic manifolds has no equivalent for topological percolation transitions on $d = 3$ hyperbolic manifolds.

8 Higher-Order Dynamics: Contagion Models

8.1 Higher-Order Contagion

The combinatorial properties of higher-order networks (including both simplicial complexes and hypergraphs) have an important effect on spreading dynamics. Contagion models on networks can be distinguished into simple contagion models and complex contagion models. In simple contagion models susceptible individuals are infected with probability λ if they are in contact with a single infected individual [5]. In models of complex contagion, susceptible individuals can be infected only if they have $\bar{k} > 1$ infected neighbors [141, 142]. While simple contagions are known to lead to continuous phase transitions, complex contagions lead to abrupt discontinuous transitions. In the framework of higher-order networks this scenario can be further enriched by the presence of many-body interactions between the nodes [14, 143–149].

The many-body interactions between the nodes can be used to formulate a higher-order contagion model [14] in which a susceptible individual has a given probability λ_1 to be infected by a single infected neighbor connected by a link (1-simplex), but has also an additional probability λ_2 to be infected by two infected neighbors belonging to the same 2-simplex. This higher-order contagion process can then be generalized to simplicial complexes of arbitrary dimensions or to hypergraphs. The introduction of additional contagion routes due to the many-body interactions in the higher-order network, is responsible for the emergence of a discontinuous transition and a region of bistability of the dynamical state of the higher-order network. In the bistability phase there is a co-existence between two different possibilities for the global spread of the contagion: the first possible dynamical outcome of the process is that the contagion does not spread, the second possible outcome is that the contagion affects a finite fraction of the nodes of the higher-order network.

Higher-order networks have also been used recently to study disease spreading. In particular, hypergraphs can be used to model co-location temporal networks in which is possible to modulate the distribution of exposure time [144]. In this context it has been shown that a bursty exposure can speed up the spreading and lower the epidemic threshold of the model.

Interestingly, contagion processes are not only affected by the combinatorial properties of the underlying higher-order contact network but are also strongly affected by its underlying topology and geometry, as shown by Ref. [150]. However, here, due to space limitation, we discuss only the effect that the combinatorial properties of higher-order networks have on spreading dynamics.

8.2 Higher-Order Contagion on the Configuration Model of Hypergraphs

In this section we consider the higher-order contagion model on the configuration model of hypergraphs described in Section 2.7. In the higher-order contagion model [14, 143] we assume that every node can be in two possible states: susceptible or infected. A susceptible individual becomes infected with rate λ_m if it belongs to a hyperedge including m other nodes which are all infected. Therefore this process describes complex contagion occurring on the hyperedge (see Figure 52 for a schematic representation of the process). An infected individual can recover and become susceptible with rate μ. If we indicate with $\rho_r(t)$ the probability that node r is infected at time t, the dynamical equation for $\rho_r(t)$ reads

$$\frac{d\rho_r}{dt} = (1 - \rho_r) \sum_{m=1}^{d} \lambda_m \Phi_m - \mu\rho_r, \tag{8.1}$$

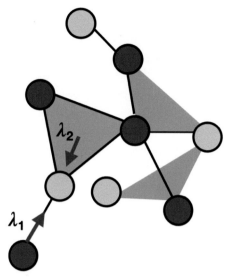

Figure 52 Schematic representation of the higher-order contagion model. Red nodes indicate infected nodes, grey nodes represent susceptible nodes. A susceptible node can be infected by a single infected node connected by a link with rate λ_1. Moreover, with rate λ_2 a susceptible node can be infected by a pair of infected nodes connected to the susceptible node by the same hyperedge of size 3. In general, a susceptible node is infected by a m-tuple of nodes connected to the susceptible node through the same $(m + 1)$-hyperedge with rate λ_m.

where Φ_m, in the mean-field approximation, is given by

$$\Phi_m = \sum_{\alpha=[v_1,v_2,\ldots v_m]\in \mathcal{Q}_m(N)} a^{[m]}_{rv_1 v_2,\ldots,v_m} \prod_{j=1}^{m} \rho_{v_j}. \tag{8.2}$$

These equations can be studied in the annealed approximation, which consists in assuming that the probability that a node is infected only depends on its generalized degree $\mathbf{k}_r = \mathbf{k}$,

$$\rho_r \to \rho_{\mathbf{k}}, \tag{8.3}$$

and approximating the adjacency tensor with its expectation in the ensemble of hypergraphs introduced in Section 2.7. Therefore, by using Eq. (2.37) we make the following additional annealed approximation

$$a^{[m]}_{\alpha} \to p^{[m]}_{\alpha} = m! k_{m,0}([r]) \prod_{j=1}^{d} \left(\frac{k_{m,0}([v_j])}{\langle k_{m,0} \rangle N} \right). \tag{8.4}$$

We can then approximate Φ_m as

$$\Phi_m = \sum_{\alpha=[v_1,v_2,\ldots v_m]\in Q_m(N)} p^{[m]}_{rv_1v_2,\ldots,v_m} \prod_{j=1}^{m} \rho_{\mathbf{k}_j}, \tag{8.5}$$

where we have indicated by $\mathbf{k}_j$ the vector of generalized degrees of node v_j. By expressing the sum over all simplices α as a sum over the m-tuples of nodes, i.e.

$$\sum_{\alpha\in Q_m(N)} \rightarrow \frac{1}{m!} \sum_{v_1=1}^{N} \sum_{v_2=1}^{N} \cdots \sum_{v_m=1}^{N} \tag{8.6}$$

and using Eq. (8.4) to express $p^{[m]}_{rv_1v_2,\ldots,v_m}$ we get

$$\Phi_m = k_{m,0}\Theta_m \tag{8.7}$$

with Θ_m given by

$$\Theta_m = \left[\sum_{\mathbf{k}} \frac{k_{m,0}}{\langle k_{m,0}\rangle} P(\mathbf{k})\rho_{\mathbf{k}} \right]^m. \tag{8.8}$$

Therefore, in the annealed approximation, the higher-order contagion dynamics are captured by the dynamical system of equations

$$\frac{d\rho_{\mathbf{k}}}{dt} = (1-\rho_{\mathbf{k}}) \sum_{m=1}^{d} \lambda_m k_{m,0}\Theta_m - \rho_{\mathbf{k}}, \tag{8.9}$$

where we have assumed, without loss of generality, that $\mu = 1$, as this scenario can be achieved by a simple rescaling of the time $t \rightarrow t\mu$ and of the parameters $\lambda_m \rightarrow \lambda_m/\mu$. The stationary state of the higher-order contagion model described by Eq. (8.9) is given by

$$\rho_{\mathbf{k}} = \frac{\sum_m \lambda_m k_{m,0}\Theta_m}{1 + \sum_{m=1}^{d} \lambda_m k_{m,0}\Theta_m}, \tag{8.10}$$

where the order parameters Θ_m defined in Eq. (8.8) satisfies the self-consistent set of equations

$$\Theta_m = \left[\sum_{\mathbf{k}} P(\mathbf{k}) \frac{k_{m,0}}{\langle k_{m,0}\rangle} \frac{\sum_{m'} \lambda_{m'} k_{m',0}\Theta_{m'}}{1 + \sum_{m'=1}^{d} \lambda_{m'} k_{m',0}\Theta_{m'}} \right]^m = F_m(\mathbf{\Theta}), \tag{8.11}$$

where we have indicated with $\mathbf{\Theta}$ the vector $\mathbf{\Theta} = (\Theta_1, \Theta_2, \ldots, \Theta_d)$. The trivial, contagion-free solution is the solution $\mathbf{\Theta} = \mathbf{0}$. However, the system of equations also admits non-trivial solutions $\mathbf{\Theta} \neq \mathbf{0}$. While in simple contagion processes corresponding to the case in which $\lambda_1 = \lambda$ and $\lambda_m = 0$ for $m > 1$, the non-trivial solution emerges continuously at the epidemic threshold $\lambda = \lambda_{1c}$, in higher-order contagions the non-trivial solution can emerge abruptly at a

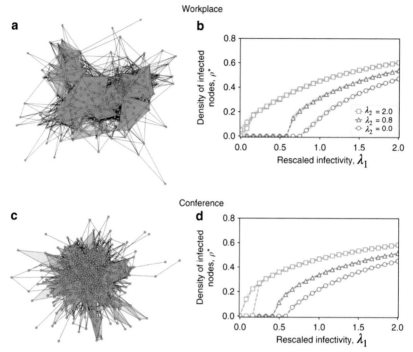

Figure 53 The higher-order contagion model simulated over the clique complex of two real social networks displays a discontinuous (for values of $\lambda_2 > \lambda_2^\star$) or a continuous (for values of $\lambda_2 \leq \lambda_2^\star$) phase transition as a function of λ_1.

Source: Reprinted from [14].

discontinuous transition and it is also possible to observe a region of bistability of the contagion processes with two states co-existing for a given set of $\lambda = (\lambda_1, \lambda_2, \ldots, \lambda_d)$: a contagion-free state and a state in which the contagion spreads. This different possible phenomenology has also been tested on real social network data in Ref. [14] (see Figure 53).

The phase diagram of the higher-order contagion process contains a contagion-free phase, an infection phase without bistability and a bistable phase. The infection phase without bistability is the region of the phase diagram characterized by the infectivities λ for which the contagion-free solution $\Theta = 0$ of Eq. (8.11) is no longer stable. To study the trivial solution $\Theta = 0$, we consider the largest eigenvalue Λ of the Jacobian matrix $\mathbf{J}$ of Eqs. (8.11) of elements given by

$$J_{mm'} = m \left(\Theta_m\right)^{1-1/m} \sum_{\mathbf{k}} P(\mathbf{k}) \lambda_{m'} \frac{k_{m,0} k_{m',0}}{\langle k_{m,0} \rangle} \frac{1}{(1 + \sum_{m''=1}^{d} \lambda_{m''} k_{m'',0} \Theta_{m''})^2}.$$

(8.12)

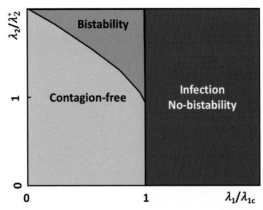

Figure 54 A schematic representation of the phase diagram of a contagion process on a hypergraph configuration model including pariwise interactions and 3-body interactions only. For $\lambda_1 > \lambda_{1c}$ the contagion process is in an infection phase, without bistability. For $\lambda < \lambda_{1c}$ it is possible to find two phases, a contagion-free phase and a phase with bistability, depending on the value of λ_2. For a quantitative phase diagram of a specific hypergraph see Ref. [143].

The trivial solution $\Theta = 0$ is unstable if and only if

$$1 < \Lambda|_{\Theta=0}. \tag{8.13}$$

Since the Jacobian matrix $\mathbf{J}$ calculated at $\Theta = 0$ has only one non-zero eigenvalue

$$\Lambda = \lambda_1 \frac{\langle k_{1,0}^2 \rangle}{\langle k_{1,0} \rangle}, \tag{8.14}$$

for the case in which we have only links and triangles, the condition for being in the infection region of the phase diagram without bistability is simply

$$\lambda_1 \geq \lambda_{1c} = \frac{\langle k_{1,0} \rangle}{\langle k_{1,0}^2 \rangle}, \tag{8.15}$$

independent of the values of λ_2 with $m > 1$. Note that the critical value λ_{1c} is equal to the epidemic threshold of the simple contagion model taking place in the network formed by hyperedges of size 2 (the links) present in the hypergraph.

In order to identify the bistable region of the phase diagram we need to investigate when the non-trivial solution $\Theta \neq 0$ of Eqs. (8.11) is stable, as long as $\lambda_1 < \lambda_{1c}$. This will allow us to predict the position of the discontinuous phase transition occurring in the higher-order contagion model. The discontinuous phase transition can be predicted to occur for the values of the infectivity

$\lambda = \lambda^\star$ for which the self-consistent Eqs. (8.11) are satisfied together with the condition

$$1 = \Lambda|_{\Theta = F(\Theta)} \qquad (8.16)$$

where Λ is the maximum eigenvalue of the Jacobian matrix $\mathbf{J}$ calculated at the non-trivial solution. This theoretical framework leads, in the case of a hypergraph with up to 3-body interactions, to a phase diagram schematically represented in Figure 54, as revealed in Ref. [143].

9 Outlook

Complex systems ranging from the brain to society are characterized by having many-body interactions. Taking into account the many-body interactions present in complex systems is key to unveiling the fundamental rules determining the relation between their underlying higher-order network structure and their function.

Already at the structural level, higher-order networks are much richer than simple networks only encoding their pairwise interactions. Therefore new tools are necessary for extracting relevant information from higher-order network data. These tools need to use not only the rich combinatorial structure of higher-order networks but also their intrinsic simplicial topology and geometry. In this respect TDA, originally developed for point clouds, is a very powerful approach that has been shown to reveal information that cannot be inferred by other statistical methods. The field is developing fast and I am sure that this line of research has the potential to be further enriched by the interaction with network science.

Simplicial geometry provides further important resources for understanding the underlying architecture of simplicial complexes. In particular, the relation between simplicial geometry and the spectral properties of the Laplacian matrices, including the recent finding that a simplicial complex can be characterized by a vector of spectral dimensions, opens a completely new perspective on the study of diffusion in simplicial complexes.

Inspired by the problem of pre-geometry initially formulated in quantum gravity, here we have discussed recent progress on the long-standing problem of emergent geometry. In particular, the Network Geometry with Flavor (NGF) is a non-equilibrium model of a growing simplicial complex which is able to generate hyperbolic simplicial complexes evolving thanks to combinatorial rules that make no use of the natural hyperbolic embedding of the generated simplicial complexes. Research on emergent geometry is likely to develop further in the near future, revealing important combinatorial relations between emergent simplicial geometry and complexity.

Finally, the field of higher-order dynamics is becoming one of the most active areas of network theory, and it is evolving at a very fast pace. The field is very promising and I have no doubt that it will be key to gain a deeper understanding of the very important relation between higher-order network structure and dynamics. Already the insights that we have obtained so far are transforming the way we look at complex systems. The formulation of dynamical models for topological signals and their dependence on the spectral properties of simplicial complexes is a clear sign that simplicial network topology and simplicial network geometry play a fundamental role in determining higher-order dynamics. This very novel point of view will be very fruitful and beneficial in making important steps forward in understanding complex systems and their interdisciplinary applications.

By highlighting the role that combinatorial, topological and geometrical properties of simplicial complexes have on their dynamics, this Element aims to provide a good starting point for stimulating young researchers to formulate their own new ideas on related subjects.

Appendix A

Maximum Entropy Ensembles of Simplicial Complexes

A1 Microcanonical and Canonical Ensembles of Simplicial Complexes

In this appendix we will provide additional information on maximum entropy ensembles of simplicial complexes. The material presented here is based on Ref. [12]. Let us indicate with $\Omega_{\mathcal{K}}$ the set of all possible simplicial complexes. Conjugated microcanonical and canonical ensembles of simplicial complexes are maximum entropy ensembles that assign to each simplicial complex $\mathcal{K}$ a probability $P(\mathcal{K})$ found by maximizing the entropy S of the ensemble

$$S = - \sum_{\mathcal{K} \in \Omega_{\mathcal{K}}} P(\mathcal{K}) \ln P(\mathcal{K}), \tag{A-1}$$

where for $P(\mathcal{K}) = 0$ we use the convention $0 \ln 0 = 0$, given a set of conjugated hard and soft constraints. In particular, the hard constraints

$$F_\mu(\mathcal{K}) = C_\mu, \tag{A-2}$$

with $\mu =\in \{1, 2, \ldots \hat{P}\}$ are imposed in the case of the microcanonical ensemble and the conjugated soft constraints

$$\sum_{\mathcal{K} \in \Omega_{\mathcal{K}}} F_\mu(\mathcal{K}) P(\mathcal{K}) = C_\mu, \tag{A-3}$$

with $\mu =\in \{1, 2, \ldots \hat{P}\}$ imposed in the case of the canonical ensemble. We now derive the expression for the probably $P(\mathcal{K})$ of simplicial complexes in the microcanonical ensemble. The hard constraints of the microcanonical ensemble are imposed by assigning non-zero probability $P(\mathcal{K}) > 0$ only to simplicial complexes in the set $\mathcal{A}_{\mathcal{K}}$ of simplicial complexes satisfying the hard constraints given by Eq. (A-2). Therefore the entropy Σ of the microcanonical ensemble can be also written as

$$\Sigma = - \sum_{\mathcal{K} \in \mathcal{A}_{\mathcal{K}}} P(\mathcal{K}) \ln P(\mathcal{K}). \tag{A-4}$$

Maximizing this expression of the entropy under the normalization condition for $P(\mathcal{K})$, its is straightforward to find that $P(\mathcal{K})$ is the uniform distribution over all simplicial complexes satisfying the hard constraints, i.e. $\mathcal{K} \in \mathcal{A}_{\mathcal{K}}$. Therefore it follows immediately that $P(\mathcal{K})$ is given by Eq. (2.18) and the entropy of the ensemble is given by Eq. (2.19). In other words, the entropy Σ of the micorcanonical ensemble is given by the logarithm of the total number $\mathcal{N}$ of simplicial complexes satisfying the hard constraints, i.e.

$$\Sigma = \ln \mathcal{N} = \ln \left[\sum_{\mathcal{K} \in \Omega_{\mathcal{K}}} \prod_{\mu=1}^{\hat{P}} \delta(C_\mu, F_\mu(\mathcal{K})) \right]. \tag{A-5}$$

Let us now derive the expression of the probability $P(\mathcal{K}) = P_C(\mathcal{K})$ of simplicial complexes in the canonical ensemble given by the Gibbs measure

$$P_C(\mathcal{K}) = \frac{1}{Z_C} e^{-\sum_{\mu=1}^{\hat{P}} \lambda_\mu F_\mu(\mathcal{K})}. \tag{A-6}$$

Here, Z_C is the normalization sum, also called the partition function, and where the Lagrangian multipliers λ_μ are fixed by the soft constraints in Eq. (A-3). In order to derive Eq.(A-6), we maximize the entropy of the ensemble S (given by Eq.(A-1)) under the constraints given by Eq. (A-3) and the condition that $P(\mathcal{K})$ is normalized. We define a functional $\mathcal{F}$ in which we have introduced the Lagrangian multipliers λ_μ, and ν

$$\mathcal{F} = -\sum_{\mathcal{K} \in \Omega_{\mathcal{K}}} P(\mathcal{K}) \ln P(\mathcal{K}) - \sum_\mu \lambda_\mu \left[\sum_{\mathcal{K} \in \Omega_{\mathcal{K}}} P(\mathcal{K}) F_\mu(\mathcal{K}) - C_\mu \right]$$
$$- \nu \left[\sum_{\mathcal{K} \in \Omega_{\mathcal{K}}} P(\mathcal{K}) - 1 \right].$$

Performing the functional derivative, and putting the functional derivative to zero,

$$\frac{\partial \mathcal{F}}{\partial P(\mathcal{K})} = -\ln P(\mathcal{K}) - 1 - \sum_{\mu=1}^{\hat{P}} \lambda_\mu F_\mu(\mathcal{K}) - \nu = 0, \tag{A-7}$$

we obtain $P(\mathcal{K}) = P_C(\mathcal{K})$ with

$$P_C(\mathcal{K}) = e^{-\sum_{\mu=1}^{\hat{P}} \lambda_\mu F_\mu(\mathcal{K}) - \nu - 1}. \tag{A-8}$$

Imposing the normalization condition, i.e. putting $\partial \mathcal{F}/\partial \nu = 0$, we derive the expression for the Gibbs measure defined in Eq. (A-6), where

$$Z_C = e^{\nu+1} = \sum_{\mathcal{K} \in \Omega_{\mathcal{K}}} e^{-\sum_{\mu=1}^{\hat{P}} \lambda_\mu F_\mu(\mathcal{K})}, \tag{A-9}$$

and where the Lagrangian multipliers λ_μ are fixed, imposing the constraints obtained by putting $\partial \mathcal{F}/\partial \lambda_\mu = 0$, i.e.

$$\sum_{\mathcal{K}} P_C(\mathcal{K}) F_\mu(\mathcal{K}) = \sum_{\mathcal{K} \in \Omega_{\mathcal{K}}} \frac{1}{Z_C} e^{-\sum_{\mu=1}^{\hat{P}} \lambda_\mu F_\mu(\mathcal{K})} F_\mu(\mathcal{K}) = C_\mu, \tag{A-10}$$

for $1 \leq \mu \leq \hat{P}$. By inserting Eq. (A-6) into the definition of the entropy (Eq.(A-1)) we obtain the entropy of the canonical ensemble, given by

$$S = \sum_{\mu=1}^{\hat{P}} \lambda_\mu C_\mu + \ln Z_C. \tag{A-11}$$

We now want to derive Eq. (2.20) relating the entropy Σ of a micro-canonical simplicial complex ensemble with the entropy S of its conjugated canonical simplicial complex ensemble. Equation (2.20) is here rewritten for convenience:

$$\Sigma = S - \hat{\Omega}, \tag{A-12}$$

where $\hat{\Omega}$ is defined as

$$\hat{\Omega} = -\ln\left[\sum_{\mathcal{K} \in \Omega_\mathcal{K}} P_C(\mathcal{K}) \prod_{\mu=1}^{\hat{P}} \delta(C_\mu, F_\mu(\mathcal{K}))\right]. \tag{A-13}$$

By inserting the explicit expression for $P_C(\mathcal{K})$ given by Eq. (A-6) in the above definition of $\hat{\Omega}$ we obtain

$$\hat{\Omega} = -\ln\left[\frac{1}{Z_C}e^{-\sum_{\mu=1}^{\hat{P}} \lambda_\mu C_\mu} \sum_{\mathcal{K} \in \Omega_\mathcal{K}} \prod_{\mu=1}^{\hat{P}} \delta(C_\mu, F_\mu(\mathcal{K}))\right]. \tag{A-14}$$

Using the expression for the entropy S of the canonical ensemble given by Eq. (A-11),

$$\hat{\Omega} = -\ln\left[e^{-S}\mathcal{N}\right] = S - \Sigma, \tag{A-15}$$

from which Eq. (2.20) follows immediately.

A2 Canonical Ensembles of Simplicial Complexes with a Given Sequence of Generalized Degree of the Nodes

In this paragraph we consider the canonical ensemble of d-dimensional pure simplicial complexes with a given sequence of generalized degree of the nodes, which is the maximum entropy ensemble given the set of N soft constraints

$$\bar{k}_{d,0}([r]) = \sum_{\mathcal{K} \in \Omega_\mathcal{K}} \left[P(\mathcal{K})\left(\sum_{v_1 < v_2 < \ldots < v_d}^{N} a_{r,v_1,v_2,\ldots,v_d}\right)\right] \tag{A-1}$$

with $r \in \{1, 2, \ldots, N\}$ and $\mathbf{a}^{[d]} = \mathbf{a}$ indicating the adjacency tensor. According to the general results presented in the previous paragraph, the probability of a simplicial complex $\mathcal{K}$ with adjacency tensor $\mathbf{a}$ is given by the Gibbs measure

$$P(\mathbf{a}) = \frac{1}{Z}e^{-H}, \tag{A-2}$$

where the Hamiltonian H is given by

$$H = \sum_{r=1}^{N} \lambda_r \sum_{v_1 < v_2 < \ldots < v_d} a_{r v_1 v_2 \ldots v_d}. \tag{A-3}$$

It is easy to show that given the invariance of the adjacency tensor under permutation of the order of its indices, the Hamiltonian can be also written as

$$H = \sum_{\alpha \in Q_d(N)} a_\alpha \left(\sum_{r \subset \alpha} \lambda_r \right), \tag{A-4}$$

from which it follows directly that

$$p_\alpha = \sum_{\mathbf{a}} P(\mathbf{a}) a_\alpha = \frac{e^{-\sum_{r \subset \alpha} \lambda_r}}{1 + e^{-\sum_{r \subset \alpha} \lambda_r}}. \tag{A-5}$$

The Lagrangian multipliers $\{\lambda_r\}$ are fixed by the condition

$$\bar{k}_{d,0}([r]) = \sum_{\alpha | r \subset \alpha} p_\alpha. \tag{A-6}$$

By solving this set of equations in the presence of the structural cutoff, i.e. for

$$\bar{k}_{d,0}([r]) \ll K = \left[\frac{(\langle k_{d,0}([r]) \rangle N)^d}{d!} \right]^{1/(d+1)}, \tag{A-7}$$

we get the uncorrelated expression of the marginal probabilities [12]

$$p_\alpha = d! \frac{\prod_{r \subset \alpha} k_{d,0}([r])}{(\langle k_{d,0}([r]) \rangle N)^d}. \tag{A-8}$$

Appendix B

The Hodge Decomposition

According to the Hodge decomposition, the space Ω^m of all m-chains can be decomposed as

$$\Omega^m \simeq \text{img}(\mathbf{B}_{[m]}^\top) \oplus \ker(\mathbf{L}_{[m]}) \oplus \text{img}(\mathbf{B}_{[m+1]}). \qquad (\text{B-1})$$

This result implies that a non-zero eigenvector of the m-order Laplacian is either a non-zero eigenvector of $\mathbf{L}_{[m]}^{up}$ or a non-zero eigenvector of $\mathbf{L}_{[m]}^{down}$.

Here our aim is to give some simple insights that can help the intuition of this result.

Consider an eigenvector $\mathbf{v}$ of the up-Laplacian with eigenvalue $\lambda \neq 0$. This eigenvector will satisfy

$$\mathbf{L}_{[m]}^{up}\mathbf{v} = \mathbf{B}_{[m+1]}\mathbf{B}_{[m+1]}^\top\mathbf{v} = \lambda\mathbf{v}, \qquad (\text{B-2})$$

or equivalently

$$\mathbf{v} = \frac{1}{\lambda}\mathbf{B}_{[m+1]}\mathbf{B}_{[m+1]}^\top\mathbf{v}. \qquad (\text{B-3})$$

Let us apply the down-Laplacian to the eigenvector $\mathbf{v}$, i.e. let us now calculate

$$\mathbf{L}_{[m]}^{down}\mathbf{v} = \mathbf{B}_{[m]}^\top\mathbf{B}_{[m]}\mathbf{v}. \qquad (\text{B-4})$$

By substituting the expression of $\mathbf{v}$ given by Eq. (B-3) into the right-hand side of this equation, we get

$$\mathbf{L}_{[m]}^{down}\mathbf{v} = \frac{1}{\lambda}\mathbf{B}_{[m]}^\top\mathbf{B}_{[m]}\mathbf{B}_{[m+1]}\mathbf{B}_{[m+1]}^\top\mathbf{v} = \mathbf{0}, \qquad (\text{B-5})$$

where we have used the property that the boundary of the boundary is null, i.e. $\mathbf{B}_{[m]}\mathbf{B}_{[m+1]} = \mathbf{0}$. It follows that if $\mathbf{v}$ is an eigenvector associated with a non-zero eigenvalue λ of the m-order up-Laplacian then it is an eigenvector of the m-order down-Laplacian with zero eigenvalue. Therefore, given that $\mathbf{L}_{[m]} = \mathbf{L}_{[m]}^{down} + \mathbf{L}_{[m]}^{up}$, $\mathbf{v}$ is also an eigenvector of the m-order Laplacian $\mathbf{L}_{[m]}$ with the eigenvalue λ. Similarly, it can be easily shown that if $\mathbf{v}$ is an eigenvector associated with a non-zero eigenvalue λ of the m-order down-Laplacian then it is also an eigenvector of the m-order Laplacian with the same eigenvalue. Consequently the spectrum of the m-order Laplacian includes all the eigenvalues of the m-order up-Laplacian and the m-order down-Laplacian. Finally, we mention the trivial observation that the non-zero eigenvalues in the spectrum of the

m-order up-Laplacian coincides with the non-zero eigenvalues in the spectrum of the $(m + 1)$-order down-Laplacian because we have

$$\mathbf{L}_{[m]}^{up} = \mathbf{B}_{[m+1]}\mathbf{B}_{[m+1]}^{\top},$$

$$\mathbf{L}_{[m+1]}^{down} = \mathbf{B}_{[m+1]}^{\top}\mathbf{B}_{[m+1]}. \tag{B-6}$$

Appendix C
Spectral Dimension of Euclidean Lattices

In this appendix our aim is to show that for Euclidean lattices the spectral dimension d_S of the graph Laplacian is equal to the dimension d of the lattice. In order to show this result we will first derive the spectrum of the graph Laplacian of a Euclidian d-dimensional lattice, and we will investigate the density of eigenvalues in the limit $N \to \infty$.

The graph Laplacian $\mathbf{L}_{[0]}$ of a network having adjacency matrix $\mathbf{A}$ has matrix elements $[L_{[0]}]_{rs}$ given by

$$[L_{[0]}]_{rs} = \delta(r,s)k_r - A_{rs}. \tag{C-1}$$

On a square lattice of d dimensions we associate with each node $r \in \{1, 2, \ldots N\}$ a position $\mathbf{x}_r \in \mathbb{R}^d$ with $\mathbf{x}_r = (x_r^{[1]}, x_r^{[2]}, \ldots x_r^{[d]})$. We assume that the square lattice is a square box of size L with periodic boundary conditions, with $N = L^d$. Therefore $x_r^{[\alpha]} \in \{1, 2, \ldots, L\}$. The periodic boundary conditions ensure that each node has the same degree, equal to $2d$.

Since we have that the matrix A_{rs} only couples neighboring nodes in the square lattice, the eigenvectors of the Laplacian are the Fourier modes

$$u_r(\mathbf{q}) = e^{i\mathbf{q} \cdot \mathbf{x}_r} \tag{C-2}$$

associated with the wave vector $\mathbf{q} = (q^{[1]}, q^{[2]}, \ldots q^{[d]})$. Additionally, the periodic boundary conditions imply that

$$\mathbf{q} = \frac{2\pi}{L}\tilde{\mathbf{n}}, \tag{C-3}$$

with $\tilde{\mathbf{n}}$ being a vector having integer elements taking values $\tilde{n}_i \in \{0, 1, 2 \ldots (L-1)\}$. Finally, the eigenvalues λ of the Laplacian $\mathbf{L}$ associated with each eigenvector $\mathbf{u}(\mathbf{q})$ are given by

$$\lambda(\mathbf{q}) = 2\sum_{i=1}^{d}[1 - \cos q^{[i]}]. \tag{C-4}$$

For $|\mathbf{q}| \ll 1$ we can develop the eigenvalues λ of the Laplacian, obtaining

$$\lambda = 2\sum_{i=1}^{d}[1 - \cos q^{[i]}] \simeq |\mathbf{q}|^2. \tag{C-5}$$

Therefore, in this approximation the eigenvalues of the Laplacian only depend on the absolute value of the wave vector. Note that in this case, as $N \to \infty$, also $L \to \infty$ and therefore, according to Eq. (C-3), the smallest non-zero absolute value of the wave-vectors goes to zero $|\mathbf{q}|_{min} \to 0$ and therefore also $\lambda_2 \to 0$, i.e. the spectral gap closes.

Equation (C-3) implies that the number of eigenmodes $\Delta\tilde{n}$ whose wave numbers have absolute value $|\mathbf{q}| = q$ in the interval $(q, q + \Delta q)$ is given by

$$\Delta\tilde{n} = \left(\frac{L}{2\pi}\right)^d \tilde{\Omega}_d q^{d-1} \Delta q, \tag{C-6}$$

where $\tilde{\Omega}_d$ is the angular integral over the d-dimensional spheric surface.

Therefore the density of states $\rho(\lambda)$ is determined by the equation

$$\Delta\tilde{n} = \rho(\lambda)\Delta\lambda, \tag{C-7}$$

where $\Delta\tilde{n}$ is the number of eigenmodes given by Eq. (C-6). Putting all together and going in the limit $\Delta\lambda \to 0$ and $\Delta q \to 0$ we get

$$\left(\frac{L}{2\pi}\right)^d \tilde{\Omega}_d q^{d-1} dq = \rho(\lambda)d\lambda \tag{C-8}$$

where λ is given by Eq.(C-5). Therefore it follows that for $\lambda \ll 1$ the density of eigenvalues $\rho(\lambda)$ scales as

$$\rho(\lambda) \propto \lambda^{d/2-1}, \tag{C-9}$$

where d is the dimension of the lattice. This shows that for Euclidean square lattices,

$$d_S = d. \tag{C-10}$$

Appendix D
Topological Moves

Starting from a d-dimensional manifold, the topological moves describe the possible modifications that do not change the topology of the manifold. Therefore the topological moves are designed to leave topological invariants such as the Euler characteristic and the Betti numbers constant. Here we focus on topological moves that can be obtained by adding or removing a d-simplex from a manifold. The topological moves are used in different fields ranging from the study of foams and inference of financial data [151], to quantum gravity approaches [77]. Different fields also use different names for the same topological moves. Here we adopt the quantum gravity terminology and we denote the topological moves as Pachner moves. The possible Pachner moves applying to d-dimensional manifolds are indicated by two numbers $u - v$ where $v = d + 1 - u$ and $u \in \{1, 2, \ldots, d\}$. Here we discuss in detail the Pachner moves that apply to a $d = 3$-dimensional manifold. The reader will be then able to generalize the procedure to other dimensions.

Given a 3-dimensional manifold with a non-zero boundary, the topological moves describe the ways in which a tetrahedron can be glued to the boundary or can be removed from the manifold leaving the topological invariants unchanged. In Figure 55 we show how the $1 - 3$, $2 - 2$ and $3 - 1$ Pachner

(a)   **(b)**   **(c)**

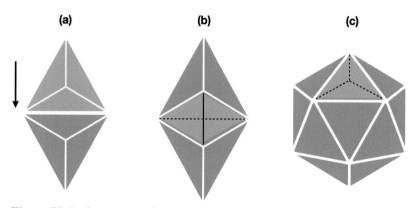

Figure 55 Pachner moves for pure simplicial complexes in dimension $d = 3$. Panel (a) Pachner move $1 - 3$: a tetrahedron is glued to a triangular face in such a way that the planar projection of the tetrahedron triangulates the original face; the reverse topological move is called Pachner move $3 - 1$. Panel (b) Pachner move $2 - 2$: a tedrahedron is glued to two existing triangular faces sharing a link; the reverse topological move is also called Pachner move $2 - 2$. Panel (c) Pachner move $3 - 1$: A tetrahedron is glued to three triangular faces in such a way that the three faces at the boundary of the manifold become one face after the move; the reverse topological move is called $1 - 3$.

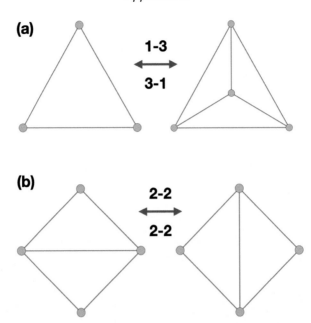

Figure 56 Pachner moves for pure simplicial complexes in dimension $d = 3$ as they appear if one considers only the faces at the boundary of the manifold. Panel (a) Pachner move $1 - 3$: a tetrahedron is glued to a triangle of the boundary in such a way that the planar projection of the tetrahedron triangulates the original triangle; the reverse topological move is called Pachner move $3 - 1$. Panel (b) Pachner move $2 - 2$: a tetrahedron is glued to two existing triangular faces connected by a link in such a way that the planar projection of the tetrahedron acts as a swap of the cordal link between the four nodes belonging to the two initial triangular faces; the reverse topological move is also called Pachner move $2 - 2$.

moves can be due to the addition of a tetrahedron glued along one, two or three faces at the boundary of the manifold. These topological moves can also be interpreted as moves applying to $d - 1$ manifolds (in our case 2-dimensional manifolds) without a boundary. These $(d - 1)$-dimensional manifolds correspond to the boundary of the d-dimensional manifolds considered in the first instance. Therefore the 3-dimensional Pachner moves described before can be considered as moves to change the tessellation of a 2-dimensional boundary, as described in Figure 56. The Pachner move $1 - 3$ and its reverse Pachner move $3 - 1$ are also known as $T1$ topological moves. The Pachner move $2 - 2$ is also known as a $T2$ topological move.

Appendix E
Emergent Preferential Attachment

In this appendix we derive Eq. (5.8) [29, 106], revealing the emergence of preferential attachment in NGFs. We will also show that in the absence of an explicit preferential attachment rule the probability $\Pi_{d,m}^{[s]}(k)$ that an m-dimensional face of generalized degree $k_{d,m} = k$ increases its generalized degree linearly with k as long as the NGF has a sufficiently high dimension d. Indeed we will show that for NGFs of dimension d and flavor s, the probability $\Pi_{d,m}^{[s]}(k)$ is given by

$$
\Pi_{d,m}^{[s]}(k) = \begin{cases} \frac{2-k}{(d-1)t} & \text{for} \quad d-m+s = 0, \\ \frac{1-s}{(d+s)t} & \text{for} \quad d-m+s = 1, \\ \frac{(d-m-1+s)k+1-s}{(d+s)t} & \text{for} \quad d-m+s > 1. \end{cases} \tag{E-1}
$$

In order to derive this expression we observe that the probability $\Pi_{d,m}^{[s]}(k)$ is the sum of the probabilities that any $(d-1)$ face $\alpha' \supseteq \alpha$ is chosen for attaching a new simplex, i.e.

$$
\begin{aligned}
\Pi_{d,m}^{[s]}(k) &= \sum_{\alpha' \in S_{d-1}(\mathcal{K})|\alpha' \supseteq \alpha} \Pi_{d,d-1}^{[s]}(\alpha') \\
&= \frac{1}{Z_t} \sum_{\alpha' \in S_{d,d-1}|\alpha' \supseteq \alpha} 1 - s + s k_{d,d-1}(\alpha'),
\end{aligned} \tag{E-2}
$$

where for $t \gg 1$ we have $Z_t = \sum_{\alpha' \in S_{d-1}(\mathcal{K})}[1 + s n_{\alpha'}] \simeq (d+s)t$. Therefore proving that Eq. (E-1) holds reduces to proving that

$$
\sum_{\alpha' \in S_{d-1}(\mathcal{K})|\alpha' \supseteq \alpha} [1 - s + s k_{d,d-1}(\alpha')] = (1-s) + (d-m-1+s)k_{d,\delta}(\alpha). \tag{E-3}
$$

In order to show that Eq. (E-3) holds, we observe that the combinatorial properties of the generalized degree (i.e. Eq. (2.3)) imply the following relation

$$
\sum_{\alpha' \in S_{d-1}(\mathcal{K}|\alpha' \supset \alpha} k_{d,d-1}(\alpha') = (d-\delta)k_{d,m}(\alpha). \tag{E-4}
$$

Additionally, for the NGF model the following relation also holds

$$
\sum_{\alpha' \in S_{d-1}(\mathcal{K})|\alpha' \supset \alpha} 1 = 1 + (d-m-1)k_{d,m}(\alpha). \tag{E-5}
$$

The left-hand side of this equation is given by the number of $(d-1)$-faces incident to the m-dimensional face α. As the NGF evolves by the addition of d-simplices, the number of $(d-1)$-faces incident to an m-dimensional face

α can be calculated by considering how many $(d-1)$-faces are added to it as the simplical network evolves. Let us assume that the face α arrives in the simplicial complex at time t' and that during the simplicial complex evolution it has acquired a generalized degree $k_{d,m}(\alpha)$. In order to calculate the left-hand side of Eq. (E-5) it is suggested to first calculate how many $(d-1)$-dimensional faces are incident to α exactly at time t', where the face is incident to a single d-simplex, and then calculate how many $(d-1)$-faces of any of the other $k_{d,m}(\alpha)-1$ d-simplices are incident to α.

At time t' the face α is incident to $d-m$ $(d-1)$-dimensional faces. In fact, a single d-dimensional simplex is incident to the face α and there are

$$\binom{d-m}{d-m-1} = d-m \text{ ways to choose the } d-(1+m) \text{ nodes of a } (d-1)\text{-face}$$

that do not belong to the m-face α out of the $d-m$ nodes of the d-simplex that do not belong to the m-face α. Similarly, one can easily see that every d-simplex that further increases the generalized degree of the m-face α contributes to the sum just by $d-m-1$.

By using Eq. (E-3), we can express $\Pi_{d,m}(\alpha)$ as Eq. (E-1).

Appendix F
Generalized Degree Distributions of NGFs (Neutral Model)

In this appendix we derive the generalized degree distribution of the NGF (neutral model) of dimension d and flavor s, using the master equation approach [29]. We have shown in the previous appendix that in a d-dimensional NGF with flavor s, the probability $\Pi_{d,m}^{[s]}(k)$ of attaching a d-dimensional simplex to an m-dimensional face with generalized degree $k_{d,m} = k$ is given by

$$\Pi_{d,m}^{[s]}(k) = \begin{cases} \frac{2-k}{(d-1)t} & \text{for} \quad d-m+s=0, \\ \frac{1-s}{(d+s)t} & \text{for} \quad d-m+s=1, \\ \frac{(d-m-1+s)k+1-s}{(d+s)t} & \text{for} \quad d-m+s>1, \end{cases} \tag{F-1}$$

as long as $d+s \neq 0$, i.e. $(d,s) \neq (1,-1)$ for which the NGF reduces to a chain. To derive the exact generalized degree distribution of the NGF, we consider the master equation for the average number of m-dimensional faces $N_{d,m}^{t,[s]}(k)$ that at time t have generalized degree $k_{d,m} = k$ in a d-dimensional simplicial NGF of flavor s. The master equation [29] for $N_{d,m}^{t,[s]}(k)$ reads

$$N_{d,m}^{t+1,[s]}(k) - N_d^{t,[s]}(k) = \Pi_{d,m}^{[s]}(k-1)N_{d,m}^{t,[s]}(k-1)[1-\delta(k,1)] - \Pi_{d,m}^{[s]}(k)N_{d,m}^{t,[s]}(k) + w_{d,m}\delta(k,1)$$

where $w_{d,m}$ indicates the number of m-dimensional faces added at each time $t > 1$ and where k is greater or equal to one, i.e. $k \geq 1$.

In the large time limit $t \gg 1$ we assume that the generalized degree approaches its asymptotic limit, i.e. $N_{d,m}^{t,[s]}(k) \simeq w_{d,m}tP_{d,m}^{[s]}(k)$, where $P_{d,m}^{[s]}(k)$ indicates the generalized degree distribution. Therefore the generalized degree distribution $P_{d,m}^{[s]}(k)$ satisfies

$$P_{d,m}^{[s]}(k) = \frac{\Pi_{d,m}^{[s]}(k-1)}{1+\Pi_{d,m}^{[s]}(k)}P_{d,m}^{[s]}(k-1)[1-\delta(k,1)] + \frac{1}{1+\Pi_{d,m}^{[s]}(k)}\delta(k,1). \tag{F-2}$$

By solving these equations we find that the generalized degree distribution depends on the dimensions d and m and on the flavor s.

- For $d-m+s=0$ (i.e. for $m=d-1$ and $s=-1$) the generalized degree distribution $P_{d,m}^{[s]}(k)$ is bimodal and is given by

$$P_{d,m}^{[s]}(k) = \begin{cases} (d-1)/d & \text{for} \quad k=1, \\ 1/d & \text{for} \quad k=2. \end{cases} \tag{F-3}$$

- For $d - m + s = 1$ (i.e. for $m = d - 2$ and $s = -1$ or $m = d - 1$ and $s = 0$) the generalized degree distribution $P_{d,m}^{[s]}(k)$ is exponential and is given by

$$P_{d,m}^{[s]}(k) = \left(\frac{d-m}{d+1}\right)^k \frac{m+1}{d-m}. \tag{F-4}$$

- For $d - m + s > 1$ (i.e. for $m \leq d - 3$ and $s = -1$ or $m \leq d - 2$ and $s = 0$ or for $d \leq d - 1$ and $s = 1$) the generalized degree distribution $P_{d,m}^{[s]}(k)$ is power-law and is given by

$$P_{d,m}^{[s]}(k) = C \frac{\Gamma[k + (1-s)/(d-m+s-1)]}{\Gamma[k+1+(d+1)/(d-m+s-1)]}, \tag{F-5}$$

with

$$C = \frac{d+s}{d-m+1-s} \frac{\Gamma[1+(d+1)/(d-m+s-1))]}{\Gamma[1+(1-s)/(d-m+s-1)]}. \tag{F-6}$$

Indeed, for large values of the generalized degrees $k_{d,m}(\alpha) = k \gg 1$ Eq. (F-5) can be approximated as

$$P_{d,m}^{[s]}(k) \simeq C k^{-\gamma_{d,m}^{[s]}}, \tag{F-7}$$

where C is a constant and where the power-law exponent $\gamma_{d,m}^{[s]}$ is given by

$$\gamma_{d,m}^{[s]} = 2 + \frac{1+m}{d-m+s-1}. \tag{F-8}$$

It follows that these power-law generalized degree distributions are scale-free, i.e. have power-law exponent $\gamma_{d,m}^{[s]}$ in the range $(2, 3]$ if and only if

$$d \geq d_c^{[m]} = 2m + 2 - s. \tag{F-9}$$

Appendix G
Apollonian and Pseudo-Fractal Simplicial Complexes

Apollonian and pseudo-fractal simplicial complexes are deterministic models of hyperbolic geometry. The Apollonian simplicial complexes are hyperbolic manifolds; the pseudo-fractal simplicial complexes are not discrete manifolds. Both structures are small-world, having an infinite Hausdorff dimension, i.e. $d_H = \infty$. Moreover, both structures are δ-hyperbolic with $\delta = 1$.

The Apollonian and the pseudo-fractal simplicial complexes are the clique complexes of Apollonian [152] and pseudo-fractal [153] networks proposed in the early days of network science. Although the original definition of these models is in a given dimension, i.e. $d = 3$ (embeddable in $d = 2$ dimensions), Apollonian and pseudo-fractal simplicial complexes can be defined in every dimension $d \geq 2$ [154].

The d-dimensional Apollonian simplicial complexes are generated deterministically starting at time $t = 1$ from a single d-dimensional simplex. At each time $t > 1$, a d-dimensional simplex is glued to every $(d - 1)$-face added at time $t - 1$. A schematic representation of an Apollonian simplicial complex of dimension $d = 2$ is shown in Figure 57(a).

The d-dimensional pseudo-fractal simplicial complexes are generated deterministically starting at time $t = 1$ from a single d-dimensional simplex. At each time $t > 1$ a d-dimensional simplex is glued to every $(d - 1)$-face of the simplicial complex. A schematic representation of a pseudo-fractal simplicial complex of dimension $d = 2$ is shown in Figure 57(b).

Both simplicial complex models display finite spectral dimensions of the graph Laplacian [155] and of higher-order up-Laplacians [88].

The Apollonian networks have a very interesting relation to Apollonian packings and Apollonian groups [156, 157].

(a) 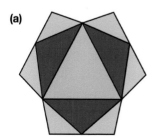 (b)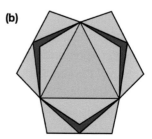

Figure 57 Schematic respresentation of Apollonian (panel a) and pseudo-fractal (panel b) simplicial complexes of dimension $d = 2$ grown up to time $t = 2$.

Source: Reprinted from [88].

References

[1] A.-L. Barabási, *Network Science*. Cambridge University Press, 2016.

[2] S. N. Dorogovtsev and J. F. Mendes, *Evolution of Networks: From Biological Nets to the Internet and WWW*. Oxford University Press, 2003.

[3] A. Barrat, M. Barthelemy and A. Vespignani, *Dynamical Processes on Complex Networks*. Cambridge University Press, 2008.

[4] F. Menczer, S. Fortunato and C. A. Davis, *A First Course in Network Science*. Cambridge University Press, 2020.

[5] S. N. Dorogovtsev, A. V. Goltsev and J. F. Mendes, "Critical phenomena in complex networks," *Reviews of Modern Physics*, vol. 80, no. 4, p. 1275, 2008.

[6] G. Bianconi, *Multilayer Networks: Structure and Function*. Oxford University Press, 2018.

[7] F. Battiston, G. Cencetti, I. Iacopini, *et al.*, "Networks beyond pairwise interactions: structure and dynamics," *Physics Reports*, vol. 874, pp. 1–92, 2020.

[8] C. Giusti, R. Ghrist and D. S. Bassett, "Two's company, three (or more) is a simplex," *Journal of Computational Neuroscience*, vol. 41, no. 1, pp. 1–14, 2016.

[9] G. Bianconi, "Interdisciplinary and physics challenges of network theory," *EPL (Europhysics Letters)*, vol. 111, no. 5, p. 56001, 2015.

[10] L. Torres, A. S. Blevins, D. S. Bassett and T. Eliassi-Rad, "The why, how, and when of representations for complex systems," *arXiv preprint arXiv:2006.02870*, 2020.

[11] V. Salnikov, D. Cassese and R. Lambiotte, "Simplicial complexes and complex systems," *European Journal of Physics*, vol. 40, no. 1, p. 014001, 2018.

[12] O. T. Courtney and G. Bianconi, "Generalized network structures: the configuration model and the canonical ensemble of simplicial complexes," *Physical Review E*, vol. 93, no. 6, p. 062311, 2016.

[13] P. Skardal and A. Arenas, "Abrupt desynchronization and extensive multistability in globally coupled oscillator simplexes," *Physical Review Letters*, vol. 122, no. 24, p. 248301, 2019.

[14] I. Iacopini, G. Petri, A. Barrat and V. Latora, "Simplicial models of social contagion," *Nature Communications*, vol. 10, no. 1, pp. 1–9, 2019.

[15] R. W. Ghrist, *Elementary Applied Topology*. Createspace Seattle, 2014, vol. 1.

[16] R. Ghrist, "Barcodes: the persistent topology of data," *Bulletin of the American Mathematical Society*, vol. 45, no. 1, pp. 61–75, 2008.

[17] M. Kahle, "Topology of random simplicial complexes: a survey," *AMS Contemporary Mathematics*, vol. 620, pp. 201–222, 2014.

[18] N. Otter, M. A. Porter, U. Tillmann, P. Grindrod and H. A. Harrington, "A roadmap for the computation of persistent homology," *EPJ Data Science*, vol. 6, no. 1, p. 17, 2017.

[19] H. Edelsbrunner, *A Short Course in Computational Geometry and Topology*. Springer, 2014.

[20] G. Petri, P. Expert, F. Turkheimer, *et al.*, "Homological scaffolds of brain functional networks," *Journal of The Royal Society Interface*, vol. 11, no. 101, p. 20140873, 2014.

[21] G. Petri, M. Scolamiero, I. Donato and F. Vaccarino, "Topological strata of weighted complex networks," *PloS One*, vol. 8, no. 6, p. e66506, 2013.

[22] M. Saggar, O. Sporns, J. Gonzalez-Castillo, *et al.* "Towards a new approach to reveal dynamical organization of the brain using topological data analysis," *Nature Communications*, vol. 9, no. 1, pp. 1–14, 2018.

[23] M. W. Reimann, M. Nolte, M. Scolamiero, *et al.*, "Cliques of neurons bound into cavities provide a missing link between structure and function," *Frontiers in Computational Neuroscience*, vol. 11, p. 48, 2017.

[24] A. P. Kartun-Giles and G. Bianconi, "Beyond the clustering coefficient: a topological analysis of node neighbourhoods in complex networks," *Chaos, Solitons and Fractals: X*, vol. 1, p. 100004, 2019.

[25] A. R. Benson, D. F. Gleich and J. Leskovec, "Higher-order organization of complex networks," *Science*, vol. 353, no. 6295, pp. 163–166, 2016.

[26] G. Palla, I. Derényi, I. Farkas and T. Vicsek, "Uncovering the overlapping community structure of complex networks in nature and society," *Nature*, vol. 435, no. 7043, pp. 814–818, 2005.

[27] A. P. Millán, J. J. Torres and G. Bianconi, "Explosive higher-order Kuramoto dynamics on simplicial complexes," *Physical Review Letters*, vol. 124, no. 21, p. 218301, 2020.

[28] S. Barbarossa and S. Sardellitti, "Topological signal processing over simplicial complexes," *IEEE Transactions on Signal Processing*, vol. 68, pp. 2992–3007, 2020.

[29] G. Bianconi and C. Rahmede, "Network geometry with flavor: from complexity to quantum geometry," *Physical Review E*, vol. 93, no. 3, p. 032315, 2016.

[30] G. Bianconi and C. Rahmede, "Emergent hyperbolic geometry," *Scientific Reports*, vol. 7, p. 41974, 2017.

[31] Z. Wu, G. Menichetti, C. Rahmede and G. Bianconi, "Emergent complex network geometry," *Scientific Reports*, vol. 5, p. 10073, 2015.

[32] D. Mulder and G. Bianconi, "Network geometry and complexity," *Journal of Statistical Physics*, vol. 173, no. 3–4, pp. 783–805, 2018.

[33] J. J. Torres and G. Bianconi, "Simplicial complexes: higher-order spectral dimension and dynamics," *Journal of Physics: Complexity*, vol. 1, no. 1, p. 015002, 2020.

[34] R. Burioni and D. Cassi, "Random walks on graphs: ideas, techniques and results," *Journal of Physics A: Mathematical and General*, vol. 38, no. 8, p. R45, 2005.

[35] A. P. Millán, J. J. Torres and G. Bianconi, "Complex network geometry and frustrated synchronization," *Scientific Reports*, vol. 8, no. 1, pp. 1–10, 2018.

[36] A. P. Millán, J. J. Torres and G. Bianconi, "Synchronization in network geometries with finite spectral dimension," *Physical Review E*, vol. 99, no. 2, p. 022307, 2019.

[37] G. Bianconi and R. M. Ziff, "Topological percolation on hyperbolic simplicial complexes," *Physical Review E*, vol. 98, no. 5, p. 052308, 2018.

[38] N. Cinardi, A. Rapisarda and G. Bianconi, "Quantum statistics in network geometry with fractional flavor," *Journal of Statistical Mechanics: Theory and Experiment*, vol. 2019, no. 10, p. 103403, 2019.

[39] G. Bianconi and C. Rahmede, "Complex quantum network manifolds in dimension d > 2 are scale-free," *Scientific Reports*, vol. 5, no. 1, pp. 1–10, 2015.

[40] A. Patania, G. Petri and F. Vaccarino, "The shape of collaborations," *EPJ Data Science*, vol. 6, no. 1, p. 18, 2017.

[41] B. Bollobás and B. Béla, *Random Graphs*. Cambridge University Press, 2001, no. 73.

[42] G. Bianconi and M. Marsili, "Emergence of large cliques in random scale-free networks," *EPL (Europhysics Letters)*, vol. 74, no. 4, p. 740, 2006.

[43] D. J. C MacKay, *Information Theory, Inference and Learning Algorithms*. Cambridge University Press, 2003.

[44] T. M. Cover, *Elements of Information Theory*. John Wiley & Sons, 1999.

[45] K. Anand and G. Bianconi, "Entropy measures for networks: toward an information theory of complex topologies," *Physical Review E*, vol. 80, no. 4, p. 045102, 2009.

[46] M. Kardar, *Statistical Physics of Particles*. Cambridge University Press, 2007.

[47] K. Anand and G. Bianconi, "Gibbs entropy of network ensembles by cavity methods," *Physical Review E*, vol. 82, no. 1, p. 011116, 2010.

[48] A. Costa and M. Farber, "Random simplicial complexes," in *Configuration Spaces*, F. Callegaro, F. Cohen, C. De Concini, E. M. Feichtner, G. Gaiffi, M. Salvetti (Eds.). Springer, 2016, pp. 129–153.

[49] K. Zuev, O. Eisenberg and D. Krioukov, "Exponential random simplicial complexes," *Journal of Physics A: Mathematical and Theoretical*, vol. 48, no. 46, p. 465002, 2015.

[50] Repository for higher-order network codes. [Online]. Available: https://github.com/ginestrab

[51] E. A. Bender and E. R. Canfield, "The asymptotic number of labeled graphs with given degree sequences," *Journal of Combinatorial Theory, Series A*, vol. 24, no. 3, pp. 296–307, 1978.

[52] G. Ghoshal, V. Zlatić, G. Caldarelli and M. E. Newman, "Random hypergraphs and their applications," *Physical Review E*, vol. 79, no. 6, p. 066118, 2009.

[53] A. E. Wegner and S. Olhede, "Atomic subgraphs and the statistical mechanics of networks," *Physical Review E*, vol. 103, no. 04, p. 042311, 2021.

[54] F. Klimm, C. M. Deane and G. Reinert, "Hypergraphs for predicting essential genes using multiprotein complex data," *bioRxiv*, 2020.

[55] H. Sun and G. Bianconi, "Higher-order percolation processes on multiplex hypergraphs," *arXiv preprint arXiv:2104.05457*, 2021.

[56] K. Zhao, J. Stehlé, G. Bianconi and A. Barrat, "Social network dynamics of face-to-face interactions," *Physical Review E*, vol. 83, no. 5, p. 056109, 2011.

[57] K. Zhao, M. Karsai and G. Bianconi, "Entropy of dynamical social networks," *PloS One*, vol. 6, no. 12, p. e28116, 2011.

[58] G. Petri and A. Barrat, "Simplicial activity driven model," *Physical Review Letters*, vol. 121, no. 22, p. 228301, 2018.

[59] P. Holme and J. Saramäki, "Temporal networks," *Physics Reports*, vol. 519, no. 3, pp. 97–125, 2012.

[60] M. Karsai, K. Kaski, A.-L. Barabási and J. Kertész, "Universal features of correlated bursty behaviour," *Scientific Reports*, vol. 2, p. 397, 2012.

[61] M. Karsai, H.-H. Jo, K. Kaski, *et al.*, *Bursty Human Dynamics*. Springer, 2018.

[62] C. Cattuto, W. Van den Broeck, A. Barrat, *et al.*, "Dynamics of person-to-person interactions from distributed rfid sensor networks," *PloS One*, vol. 5, no. 7, p. e11596, 2010.

[63] J. Stehlé, A. Barrat and G. Bianconi, "Dynamical and bursty interactions in social networks," *Physical Review E*, vol. 81, no. 3, p. 035101, 2010.

[64] G. Cencetti, F. Battiston, B. Lepri and M. Karsai, "Temporal properties of higher-order interactions in social networks," *arXiv preprint arXiv:2010.03404*, 2020.

[65] Link to SocioPattern project webpage. [Online]. Available: http://www.sociopatterns.org/

[66] J. Jost, *Mathematical Concepts*. Springer, 2015.

[67] S. H. Lee, M. D. Fricker and M. A. Porter, "Mesoscale analyses of fungal networks as an approach for quantifying phenotypic traits," *Journal of Complex Networks*, Apr 2016. [Online]. Available: https://doi.org/10.1093/comnet/cnv034

[68] Link to license. [Online]. Available: https://creativecommons.org/licenses/by/4.0/

[69] D. J. Watts and S. H. Strogatz, "Collective dynamics of 'small-world' networks," *Nature*, vol. 393, no. 6684, pp. 440–442, 1998.

[70] A. Patania, F. Vaccarino and G. Petri, "Topological analysis of data," *EPJ Data Science*, vol. 6, no. 1, pp. 1–6, 2017.

[71] G. Bianconi, R. K. Darst, J. Iacovacci and S. Fortunato, "Triadic closure as a basic generating mechanism of communities in complex networks," *Physical Review E*, vol. 90, no. 4, p. 042806, 2014.

[72] Y. Ollivier, "Ricci curvature of metric spaces," *Comptes Rendus Mathematique*, vol. 345, no. 11, pp. 643–646, 2007.

[73] F. Bauer, J. Jost and S. Liu, "Ollivier-Ricci curvature and the spectrum of the normalized graph Laplace operator," *arXiv preprint arXiv:1105.3803*, 2011.

[74] J. Jost and S. Liu, "Ollivier's Ricci curvature, local clustering and curvature-dimension inequalities on graphs," *Discrete and Computational Geometry*, vol. 51, no. 2, pp. 300–322, 2014.

[75] R. Sreejith, K. Mohanraj, J. Jost, E. Saucan and A. Samal, "Forman curvature for complex networks," *Journal of Statistical Mechanics: Theory and Experiment*, vol. 2016, no. 6, p. 063206, 2016.

[76] T. Regge, "General relativity without coordinates," *Il Nuovo Cimento (1955–1965)*, vol. 19, no. 3, pp. 558–571, 1961.

[77] B. Dittrich, L. Freidel and S. Speziale, "Linearized dynamics from the 4-simplex Regge action," *Physical Review D*, vol. 76, no. 10, p. 104020, 2007.

[78] J. Ambjørn, J. Jurkiewicz and R. Loll, "Emergence of a 4d world from causal quantum gravity," *Physical Review Letters*, vol. 93, no. 13, p. 131301, 2004.

[79] M. Gromov, "Hyperbolic groups," in *Essays in Group Theory*, S. M. Gersten (Ed.). Springer, 1987, pp. 75–263.

[80] E. Jonckheere, P. Lohsoonthorn and F. Bonahon, "Scaled Gromov hyperbolic graphs," *Journal of Graph Theory*, vol. 57, no. 2, pp. 157–180, 2008.

[81] R. Albert, B. DasGupta and N. Mobasheri, "Topological implications of negative curvature for biological and social networks," *Physical Review E*, vol. 89, no. 3, p. 032811, 2014.

[82] W. S. Kennedy, O. Narayan and I. Saniee, "On the hyperbolicity of large-scale networks," *arXiv preprint arXiv:1307.0031*, 2013.

[83] G. Calcagni, A. Eichhorn and F. Saueressig, "Probing the quantum nature of spacetime by diffusion," *Physical Review D*, vol. 87, no. 12, p. 124028, 2013.

[84] D. Benedetti and J. Henson, "Spectral geometry as a probe of quantum spacetime," *Physical Review D*, vol. 80, no. 12, p. 124036, 2009.

[85] T. Jonsson and J. F. Wheater, "The spectral dimension of the branched polymer phase of two-dimensional quantum gravity," *Nuclear Physics B*, vol. 515, no. 3, pp. 549–574, 1998.

[86] B. Durhuus, T. Jonsson and J. F. Wheater, "The spectral dimension of generic trees," *Journal of Statistical Physics*, vol. 128, no. 5, pp. 1237–1260, 2007.

[87] R. Burioni and D. Cassi, "Universal properties of spectral dimension," *Physical Review Letters*, vol. 76, no. 7, p. 1091–1093, 1996.

[88] M. Reitz and G. Bianconi, "The higher-order spectrum of simplicial complexes: a renormalization group approach," *Journal of Physics A: Mathematical and Theoretical*, vol. 53, p. 295001, 2020.

[89] V. J. Wedeen, D. L. Rosene, R. Wang, *et al.*, "The geometric structure of the brain fiber pathways," *Science*, vol. 335, no. 6076, pp. 1628–1634, 2012.

[90] E. Katifori, G. J. Szöllősi and M. O. Magnasco, "Damage and fluctuations induce loops in optimal transport networks," *Physical Review Letters*, vol. 104, no. 4, p. 048704, 2010.

[91] J. W. Rocks, A. J. Liu and E. Katifori, "Hidden topological structure of flow network functionality," *Physical Review Letters*, vol. 126, no. 2, p. 028102, 2021.

[92]   M. Á. Serrano, M. Boguná and F. Sagués, "Uncovering the hidden geometry behind metabolic networks," *Molecular Biosystems*, vol. 8, no. 3, pp. 843–850, 2012.

[93]   F. Radicchi, D. Krioukov, H. Hartle and G. Bianconi, "Classical information theory of networks," *Journal of Physics: Complexity*, vol. 1, no. 2, p. 025001, 2020.

[94]   R. Penrose, "On the nature of quantum geometry," *Magic Without Magic*, J. R. Klauder (Ed.). W. H. Freeman & Co. Ltd. pp. 333–354, 1972.

[95]   L. Smolin, *The Life of the Cosmos*. Oxford University Press, 1999.

[96]   G. Bianconi and A.-L. Barabási, "Bose–Einstein condensation in complex networks," *Physical Review Letters*, vol. 86, no. 24, p. 5632, 2001.

[97]   N. Fountoulakis, T. Iyer, C. Mailler and H. Sulzbach, "Dynamical models for random simplicial complexes," *arXiv preprint arXiv:1910.12715*, 2019.

[98]   A.-L. Barabási and R. Albert, "Emergence of scaling in random networks," *Science*, vol. 286, no. 5439, pp. 509–512, 1999.

[99]   S. N. Dorogovtsev, J. F. Mendes and A. N. Samukhin, "Size-dependent degree distribution of a scale-free growing network," *Physical Review E*, vol. 63, no. 6, p. 062101, 2001.

[100] J. Nokkala, J. Piilo and G. Bianconi, "Probing the spectral dimension of quantum network geometries," *Journal of Physics: Complexity*, vol. 2, no. 1, p. 015001, 2020.

[101] D. C. da Silva, G. Bianconi, R. A. da Costa, S. N. Dorogovtsev and J. F. Mendes, "Complex network view of evolving manifolds," *Physical Review E*, vol. 97, no. 3, p. 032316, 2018.

[102] M. Šuvakov, M. Andjelković and B. Tadić, "Hidden geometries in networks arising from cooperative self-assembly," *Scientific Reports*, vol. 8, no. 1, pp. 1–10, 2018.

[103] M. M. Dankulov, B. Tadić and R. Melnik, "Spectral properties of hyperbolic nanonetworks with tunable aggregation of simplexes," *Physical Review E*, vol. 100, no. 1, p. 012309, 2019.

[104] A. P. Millán, R. Ghorbanchian, N. Defenu, F. Battiston, and G. Bianconi, "Local topological moves determine global diffusion properties of hyperbolic higher-order networks," *arXiv preprint arXiv:2102.12885*, 2021.

[105] M. Girvan and M. E. Newman, "Community structure in social and biological networks," *Proceedings of the National Academy of Sciences*, vol. 99, no. 12, pp. 7821–7826, 2002.

[106] O. T. Courtney and G. Bianconi, "Weighted growing simplicial complexes," *Physical Review E*, vol. 95, no. 6, p. 062301, 2017.

[107] K. Kovalenko, I. Sendiña-Nadal, N. Khalil, *et al.*, "Growing scale-free simplices," *arXiv preprint arXiv:2006.12899*, 2020.

[108] G. Bianconi and A.-L. Barabási, "Competition and multiscaling in evolving networks," *EPL (Europhysics Letters)*, vol. 54, no. 4, p. 436, 2001.

[109] G. Bianconi, C. Rahmede and Z. Wu, "Complex quantum network geometries: evolution and phase transitions," *Physical Review E*, vol. 92, no. 2, p. 022815, 2015.

[110] S. Strogatz, *Sync: The Emerging Science of Spontaneous Order*. Penguin UK, 2004.

[111] Y. Kuramoto, "Self-entrainment of a population of coupled non-linear oscillators," in *International Symposium on Mathematical Problems in Theoretical Physics*, H. Araki (Ed.). Springer, 1975, pp. 420–422.

[112] S. H. Strogatz, "From Kuramoto to Crawford: exploring the onset of synchronization in populations of coupled oscillators," *Physica D: Nonlinear Phenomena*, vol. 143, no. 1–4, pp. 1–20, 2000.

[113] A. Pikovsky, M. Rosenblum and J. Kurths, *Synchronization: A Universal Concept in Nonlinear Sciences*. Cambridge University Press, 2003, no. 12.

[114] A. Arenas, A. Díaz-Guilera, J. Kurths, Y. Moreno and C. Zhou, "Synchronization in complex networks," *Physics Reports*, vol. 469, no. 3, pp. 93–153, 2008.

[115] S. Boccaletti, A. N. Pisarchik, C. I. Del Genio and A. Amann, *Synchronization: from Coupled Systems to Complex Networks*. Cambridge University Press, 2018.

[116] J. G. Restrepo, E. Ott and B. R. Hunt, "Onset of synchronization in large networks of coupled oscillators," *Physical Review E*, vol. 71, no. 3, p. 036151, 2005.

[117] A. P. Millán, J. G. Restrepo, J. J. Torres and G. Bianconi, "Geometry, topology and simplicial synchronization," *arXiv preprint arXiv:2105.00943*, 2021.

[118] C. Kuehn and C. Bick, "A universal route to explosive phenomena," *Science Advances*, vol. 7, no. 16, p. eabe3824, 2021.

[119] F. P. U. Severino, J. Ban, Q. Song, *et al.*, "The role of dimensionality in neuronal network dynamics," *Scientific Reports*, vol. 6, p. 29640, 2016.

[120] L. Gambuzza, F. Di Patti, L. Gallo, *et al.*, "The master stability function for synchronization in simplicial complexes," *arXiv preprint arXiv:2004.03913*, 2020.

[121] R. Mulas, C. Kuehn and J. Jost, "Coupled dynamics on hypergraphs: master stability of steady states and synchronization," *Physical Review E*, vol. 101, no. 6, p. 062313, 2020.

[122] Y. Zhang, V. Latora and A. E. Motter, "Unified treatment of dynamical processes on generalized networks: higher-order, multilayer, and temporal interactions," *arXiv preprint arXiv:2010.00613*, 2020.

[123] A. Salova and R. M. D'Souza, "Cluster synchronization on hypergraphs," *arXiv preprint arXiv:2101.05464, 2021*.

[124] R. Ghorbanchian, J. G. Restrepo, J. J. Torres and G. Bianconi, "Higher-order simplicial synchronization of coupled topological signals," *Communications Physics*, vol. 4, p.120, 2021.

[125] P. S. Skardal and A. Arenas, "Memory selection and information switching in oscillator networks with higher-order interactions," *Journal of Physics: Complexity*, vol. 2, no. 1, p. 015003, 2020.

[126] P. Skardal and A. Arenas, "Higher order interactions in complex networks of phase oscillators promote abrupt synchronization switching," *Communications Physics*, vol. 3, no. 1, pp. 1–6, 2020.

[127] X. Dai, K. Kovalenko, M. Molodyk, *et al.*, "D-dimensional oscillators in simplicial structures: odd and even dimensions display different synchronization scenarios," *arXiv preprint arXiv:2010.14976*, 2020.

[128] M. Lucas, G. Cencetti and F. Battiston, "Multiorder Laplacian for synchronization in higher-order networks," *Physical Review Research*, vol. 2, no. 3, p. 033410, 2020.

[129] X. Zhang, S. Boccaletti, S. Guan and Z. Liu, "Explosive synchronization in adaptive and multilayer networks," *Physical Review Letters*, vol. 114, no. 3, p. 038701, 2015.

[130] L. DeVille, "Consensus on simplicial complexes, or: The nonlinear simplicial Laplacian," *arXiv preprint arXiv:2010.07421*, 2020.

[131] I. Benjamini and O. Schramm, "Percolation in the hyperbolic plane," *Journal of the American Mathematical Society*, vol. 14, no. 2, pp. 487–507, 2001.

[132] S. Boettcher, V. Singh and R. M. Ziff, "Ordinary percolation with discontinuous transitions," *Nature Communications*, vol. 3, no. 1, pp. 1–5, 2012.

[133] I. Kryven, R. M. Ziff and G. Bianconi, "Renormalization group for link percolation on planar hyperbolic manifolds," *Physical Review E*, vol. 100, no. 2, p. 022306, 2019.

[134] G. Bianconi, I. Kryven and R. M. Ziff, "Percolation on branching simplicial and cell complexes and its relation to interdependent percolation," *Physical Review E*, vol. 100, no. 6, p. 062311, 2019.

[135] H. Sun, R. M. Ziff and G. Bianconi, "Renormalization group theory of percolation on pseudo-fractal simplicial and cell complexes," *Physical Review E*, vol. 102, p. 012308, 2020.

[136] O. Bobrowski and P. Skraba, "Homological percolation and the Euler characteristic," *Physical Review E*, vol. 101, no. 3, p. 032304, 2020.

[137] Y. Lee, J. Lee, S. M. Oh, D. Lee and B. Kahng, "Homological percolation transitions in evolving coauthorship complexes," *arXiv preprint arXiv:2010.12224*, 2020.

[138] B. C. Coutinho, A.-K. Wu, H.-J. Zhou and Y.-Y. Liu, "Covering problems and core percolations on hypergraphs," *Physical Review Letters*, vol. 124, no. 24, p. 248301, 2020.

[139] I. Amburg, J. Kleinberg and A. Benson, "Planted hitting set recovery in hypergraphs," *Journal of Physics: Complexity*, vol. 2, p. 035004, 2021.

[140] D. M. Auto, A. A. Moreira, H. J. Herrmann and J. S. Andrade Jr, "Finite-size effects for percolation on Apollonian networks," *Physical Review E*, vol. 78, no. 6, p. 066112, 2008.

[141] D. J. Watts, "A simple model of global cascades on random networks," *Proceedings of the National Academy of Science*, vol. 99, no. 9, p. 5766, 2002.

[142] M. Granovetter, "Threshold models of collective behavior," *American Journal of Sociology*, vol. 83, no. 6, p. 1420, 1978.

[143] N. W. Landry and J. G. Restrepo, "The effect of heterogeneity on hypergraph contagion models," *Chaos: An Interdisciplinary Journal of Nonlinear Science*, vol. 30, no. 10, p. 103117, 2020.

[144] G. St-Onge, H. Sun, A. Allard, L. Hébert-Dufresne and G. Bianconi, "Universal nonlinear infection kernel from heterogeneous exposure on higher-order networks." *arXiv preprint arXiv:2101.07229, 2021.*

[145] G. St-Onge, A. Allard and L. Hébert-Dufresne, "Localization, bistability and optimal seeding of contagions on higher-order networks," in *Artificial Life Conference Proceedings*. MIT Press, 2020, pp. 567–569.

[146] G. St-Onge, V. Thibeault, A. Allard, L. J. Dubé and L. Hébert-Dufresne, "Master equation analysis of mesoscopic localization in contagion dynamics on higher-order networks," *arXiv preprint arXiv:2004.10203*, 2020.

[147] G. F. de Arruda, G. Petri and Y. Moreno, "Social contagion models on hypergraphs," *Physical Review Research*, vol. 2, no. 2, p. 023032, 2020.

[148] B. Jhun, M. Jo and B. Kahng, "Simplicial SIS model in scale-free uniform hypergraph," *Journal of Statistical Mechanics: Theory and Experiment*, vol. 2019, no. 12, p. 123207, 2019.

[149] G. F. de Arruda, M. Tizzani and Y. Moreno, "Phase transitions and stability of dynamical processes on hypergraphs," *arXiv preprint arXiv:2005.10891*, 2020.

[150] D. Taylor, F. Klimm, H. A. Harrington, *et al.*, "Topological data analysis of contagion maps for examining spreading processes on networks," *Nature Communications*, vol. 6, p.7723, 2015.

[151] G. P. Massara, T. Di Matteo and T. Aste, "Network filtering for big data: Triangulated maximally filtered graph," *Journal of Complex Networks*, vol. 5, no. 2, pp. 161–178, 2016.

[152] J. S. Andrade Jr, H. J. Herrmann, R. F. Andrade and L. R. Da Silva, "Apollonian networks: simultaneously scale-free, small world, Euclidean, space filling, and with matching graphs," *Physical Review Letters*, vol. 94, no. 1, p. 018702, 2005.

[153] S. N. Dorogovtsev, A. V. Goltsev and J. F. F. Mendes, "Pseudofractal scale-free web," *Physical Review E*, vol. 65, no. 6, p. 066122, 2002.

[154] Z. Zhang, F. Comellas, G. Fertin and L. Rong, "High-dimensional Apollonian networks," *Journal of Physics A: Mathematical and General*, vol. 39, no. 8, pp. 1811–1818, 2006.

[155] G. Bianconi and S. N. Dorogovstev, "The spectral dimension of simplicial complexes: a renormalization group theory," *Journal of Statistical Mechanics: Theory and Experiment*, vol. 2020, no. 1, p. 014005, 2020.

[156] B. Söderberg, "Apollonian tiling, the Lorentz group, and regular trees," *Physical Review A*, vol. 46, no. 4, pp. 1859–1866, 1992.

[157] R. L. Graham, J. C. Lagarias, C. L. Mallows, A. R. Wilks and C. H. Yan, "Apollonian circle packings: geometry and group theory. I. The Apollonian group," *Discrete and Computational Geometry*, vol. 34, no. 4, pp. 547–585, 2005.

Acknowledgements

This work originates from the many interesting discussions and collaborations with friends and colleagues including Alain Barrat, Federico Battiston, Stefano Boccaletti, Stefan Boettcher, Gareth Baxter, Lucille Calmon, Owen Courtney, Sergey N. Dorogovstev, Rui A. da Costa, Michael Farber, Reza Ghorbanchian, Ivan Kryven, Vito Latora, Giacomo Livan, Giulia Menichetti, Ana Paula Millán, José Fernando Mendes, Giovanni Petri, Christoph Rahmede, Marcus Reitz, Juan Restrepo, Primoz Skraba, Hanlin Sun, Joaquín J. Torres, Francesco Vaccarino, Zhihao Wu, Kun Zhao and Robert M. Ziff.

I thank the Department of Mathematical Sciences Giuseppe Luigi Lagrange at Politecnico of Torino, Italy for inviting me to hold the Excellence PhD course on network theory, the School and Conference in Applied Algebraic Topology at the Hausdorff Institute in Bonn and the NetSci 2020 School for hosting my lectures on emergent geometry and higher-order dynamics. These series of lectures have been the starting point for this Element. These events could not been possible without the support of great friends and colleagues: Francesco Vaccarino, Michael Farber and Guido Caldarelli.

I really thank Guido Caldarelli and Nicholas Gibbons for inviting me to write this Element in this much needed collection of *Cambridge Elements in the Structure and Dynamics of Complex Networks*. Their great patience and encouragement was key to finishing the Element despite the many challenges found during the writing of this work.

I also thank the School of Mathematical Sciences at Queen Mary University of London that has allowed me to make this Element a reality.

Finally I am most grateful to my husband for his enthusiasm and support that has motivated me to complete this work.

About the Author

Ginestra Bianconi is Professor of Applied Mathematics at the School of Mathematical Sciences of Queen Mary University of London and Alan Turing Fellow at the Alan Turing Institute. Currently she is Editor in Chief of the Journal of Physics: Complexity, and Editor of Scientific Reports and PLoS One. She has published more than 160 articles in statistical mechanics, network theory and its interdisciplinary applications, and she is the author of the book *Multilayer Networks: Structure and Function*. For her work on the Bose–Einstein condensation in complex networks and her foundational contributions on multilayer networks she has been awarded the Network Science Fellowship by the Network Science Society. In recent years, her research interest has been focused on higher-order networks and simplicial complexes and in particular on the interplay between network topology and higher-order dynamics.

Cambridge Elements ≡

The Structure and Dynamics of Complex Networks

Guido Caldarelli
Ca' Foscari University of Venice

Guido Caldarelli is Full Professor of Theoretical Physics at Ca' Foscari University of Venice. Guido Caldarelli received his Ph.D. from SISSA, after which he held postdoctoral positions in the Department of Physics and School of Biology, University of Manchester, and the Theory of Condensed Matter Group, University of Cambridge. He also spent some time at the University of Fribourg in Switzerland, at École Normale Supérieure in Paris, and at the University of Barcelona. His main research focus is the study of networks, mostly analysis and modelling, with applications from financial networks to social systems as in the case of disinformation. He is the author of more than 200 journal publications on the subject, and three books, and is the current President of the Complex Systems Society (2018 to 2021).

About the Series

This cutting-edge new series provides authoritative and detailed coverage of the underlying theory of complex networks, specifically their structure and dynamical properties. Each Element within the series will focus upon one of three primary topics: static networks, dynamical networks and numerical/computing network resources.

Elements in the Series

Reconstructing Networks
Giulio Cimini, Rossana Mastrandrea and Tiziano Squartini

Higher-Order Networks: An Introduction to Simplicial Complexes
Ginestra Bianconi

A full series listing is available at: www.cambridge.org/SDCN

Printed in the United States
by Baker & Taylor Publisher Services